The Art and Science of Making Up Your

The Art and Science of Making Up Your Mind presents basic decision-making principles and tools to help the reader respond efficiently and wisely to everyday dilemmas.

Although most decisions are made informally (whether intuitively without deliberate thought, or based on careful reflection), over the centuries people have tried to develop systematic, scientific and structured ways in which to make decisions. Using qualitative counterparts to quantitative models, Rex Brown takes the reader through the basics, like "what is a decision" and then considers a wide variety of real-life decisions, explaining how the best judgments can be made using logical principles.

Combining multiple evaluations of the same judgment ("hybrid judgment") and exploring innovative analytical concepts (such as "ideal judgment"), this book explores and analyzes the skills needed to master the basics of non-mathematical decision making, and what should be done, using real world illustrations of decision methods.

The book is an ideal companion for students of Thinking, Reasoning and Decision-Making, and also for anyone wanting to understand how to make better judgments in their everyday lives.

Rex V. Brown's 50-year career combined research, consulting and teaching on helping people and organizations to make better decisions. He wrote five previous books on decision science and more than 80 papers and articles. Brown was a Distinguished Senior Fellow in the School of Public Policy at George Mason University. He held faculty appointments at Harvard Business School, University College London, London School of Economics and the Universities of Michigan, Carnegie-Mellon, George Mason and Cambridge. He spent 20 years as Chairman of Decision Science Consortium, Inc. in Reston, Virginia, and was a founding Council Member of the Decision Analysis Society.

This highly readable book aims to teach the reader to obtain a superior second opinion – from herself. A wealth of often provocative examples reveals the wisdom of a master of applied decision theory. – Daniel Kahneman, Nobel Prize Winner, Author of *Thinking, Fast and Slow*, Eugene Higgins Professor of Psychology Emeritus at Princeton University, USA

A trailblazing pioneer in decision education, Rex Brown has provided us with invaluable examples, tools, evidence and arguments for everyone to invest in their decision skills. – Chris Spetzler, Executive Director, Decision Education Foundation, USA

Rex Brown spent all his working life thinking about how to make decisions well. This book is the culmination of his thought. He concentrates on the essential ideas, illustrated with many practical examples, and avoiding most of the mathematics that surrounded the subject when originally formed. This excellent text is to be recommended to all who want a readable and straightforward introduction to the analysis of any decision. – Professor Stephen Watson, Life Fellow, Emmanuel College, Cambridge University, UK

Did you ever think hard about a decision and later feel like it had been a bad mistake? After decades advising others and trying to avoid making more mistakes myself, I urge you to read this book now. No psychology or statistics, just lots of pithy, how-to-do-it, real examples. You'll make better decisions easily, and you'll have fewer regrets. – Andrew Kahr, Founder: First Deposit Corporation, formerly Assoc. Prof. of Business Administration, Harvard Business School, USA

Rex Brown's The Art and Science of Making Up Your Mind *is a non-technical textbook and guidebook for how to understand and use basic principles and tools of applied decision theory to deal effectively with everyday decisions and difficult dilemmas. Rex Brown is a wise mentor and reliable companion. He guides the reader through basic questions such as 'what is a decision' and 'what is an ideal judgment.' He then draws on personal, family and friends' decisions in professional life, health, illness, and voting to elucidate how quantitative decision aids and qualitative considerations help clarify problems and lead to sound decisions. An indispensable resource for students and all decision makers.* – Professor Leon Mann AO, Professorial Fellow, School of Psychological Sciences, University of Melbourne, Australia

This gem of a book synthesizes the life's work of one of decision science's seminal thinkers. Rex Brown's passion was making the tools and thought processes of applied decision theory accessible to ordinary people. This clear and entertaining how-to manual uses examples from his own life and the lives of his family to help readers grasp how we can realize our potential for better everyday decision-making and, ultimately, greater satisfaction in our lives. — Kathryn Blackmond Laskey, Professor of Systems Engineering and Operations Research, George Mason University

The Art and Science of Making Up Your Mind

Rex V. Brown

Edited by Jonathan Baron and Karen Brown

NEW YORK AND LONDON

First published 2020
by Routledge
52 Vanderbilt Avenue, New York, NY 10017

and by Routledge
2 Park Square, Milton Park, Abingdon, Oxon, OX14 4RN

Routledge is an imprint of the Taylor & Francis Group, an informa business

© 2020 Taylor & Francis

The right of Jonathan Baron and Karen Brown to be identified as the authors of the editorial material, and of the authors for their individual chapters, has been asserted in accordance with sections 77 and 78 of the Copyright, Designs and Patents Act 1988.

All rights reserved. No part of this book may be reprinted or reproduced or utilised in any form or by any electronic, mechanical, or other means, now known or hereafter invented, including photocopying and recording, or in any information storage or retrieval system, without permission in writing from the publishers.

Trademark notice: Product or corporate names may be trademarks or registered trademarks, and are used only for identification and explanation without intent to infringe.

Library of Congress Cataloging-in-Publication Data
A catalog record for this title has been requested

ISBN: 978-1-84872-917-9 (hbk)
ISBN: 978-1-84872-657-4 (pbk)
ISBN: 978-0-367-27863-2 (ebk)

Typeset in Computer Modern by Jonathan Baron

Contents

Editors' preface	xiii

1 Introduction   1
   1.1 Potential need 1
   1.2 Distinctive features of this text 3
   1.3 Vocabulary 3
   1.4 Historical background 4
      1.4.1 ADT methods 6
      1.4.2 My own ADT evolution 6
   1.5 Looking ahead 8

2 Decision making   9
   2.1 What is a decision? 9
      2.1.1 Ideal decision 10
      2.1.2 Value of wisdom 10
   2.2 Requirements of wise action 11
      2.2.1 Accounting for all available knowledge 11
      2.2.2 Wisdom is no substitute for knowledge 12
      2.2.3 Good outcomes may not imply wise decisions 13
   2.3 Current decision practice 14
      2.3.1 The world of private decisions 14
      2.3.2 Room for improvement 14
      2.3.3 Motivation for wisdom 15
      2.3.4 Opposition to rationality 16
      2.3.5 Raw material: The contents of your mind 17

		2.3.6	Decision vs. action	17
		2.3.7	Precept vs. process	18
	2.4	Decider roles		18
		2.4.1	Personal decisions	19
		2.4.2	Civic decisions	19
		2.4.3	Civic vs. professional decisions	20
	2.5	Student end product		21
	2.6	Exercises		22
		2.6.1	Bad outcome — good decision	22
		2.6.2	Decider role	23
3	**Term project: Evaluating a policy proposal**			**25**
	3.1	Task		25
	3.2	Nature of issues to be evaluated		25
	3.3	Project activity during course		26
	3.4	Sample proposals		26
	3.5	Final report		27
		3.5.1	Suggested report sections	27
	3.6	Standing assignment after each chapter		28
4	**Qualitative decision aids**			**29**
	4.1	Types of decision aid		29
		4.1.1	Prescriptive decision analysis	29
		4.1.2	Applied decision theory	31
	4.2	Projection		31
		4.2.1	Visualizing option aftermaths	31
	4.3	Anchoring judgments		33
		4.3.1	Comparing this dilemma with past decisions	33
		4.3.2	"Expert" opinion	34
		4.3.3	Adapted decision rules	34
	4.4	Decomposing a decision into stages		35
		4.4.1	Going through "HOOPS"	35
		4.4.2	Caution — Confusing factual and value judgments	36
		4.4.3	Implementing HOOPS	37
	4.5	Exercises		37

	4.5.1	Informal reflection	37
	4.5.2	Asking the right questions	38
	4.5.3	Other exercises	38
	4.5.4	Distinguishing factual from value judgments	41

5 Hip surgery case   43

- 5.1 Preliminary evaluations of hip choice 44
  - 5.1.1 Intuition 44
  - 5.1.2 Consultation 44
  - 5.1.3 Cognitive vigilance 44
  - 5.1.4 Analogy 44
  - 5.1.5 Visualized aftermaths 45
  - 5.1.6 Interim synthesis of evaluations 45
- 5.2 Main evaluation effort 45
  - 5.2.1 Student participation in hip choice 45
  - 5.2.2 Simplifying the options 46
  - 5.2.3 Choice criteria and option impacts 47
- 5.3 Synthesis of conflicting appraisals 48
  - 5.3.1 Synthesis at the time 48
  - 5.3.2 Hindsight synthesis 49
- 5.4 Seek more information before deciding? 49
- 5.5 Post scripts 50
  - 5.5.1 Sequel 50
  - 5.5.2 Action vs. decision 50
  - 5.5.3 A later choice 50
- 5.6 Exercise 51

6 Quantitative ADT modeling   53

- 6.1 Quantitative decision reasoning 53
  - 6.1.1 Reasoned decisions 53
  - 6.1.2 Models 53
- 6.2 Applied decision theory (ADT) 54
  - 6.2.1 Normative decision theory 54
  - 6.2.2 Applied decision theory models 54
  - 6.2.3 Utility as a measure of satisfaction 55

	6.2.4 Uncertain utility	55
6.3	Quantifying HOOPS: Decision trees	56
	6.3.1 Basic decision tree mechanics: Simple example	56
	6.3.2 "Bachelor Joe" example	57
6.4	Assignments	59

7 Family case study: C-section vs. "natural" child-birth? 61
- 7.1 A baby delivery dilemma . . . 61
- 7.2 Postscript (six months later) . . . 66
  - 7.2.1 Post-postscript . . . 67
- 7.3 Technical commentary . . . 67
  - 7.3.1 Decision-making issues raised . . . 67
  - 7.3.2 Applied decision theory . . . 68
  - 7.3.3 My contribution as analyst . . . 68
- 7.4 Exercises . . . 68

8 Using ADT models 71
- 8.1 Formal vs. informal evaluations . . . 71
- 8.2 Caution . . . 72
- 8.3 Combining ADT with alternative evaluations . . . 73

9 A civic case: Voting for president 75
- 9.1 Leah's vote . . . 75
- 9.2 Abe's vote . . . 78
- 9.3 Exercise . . . 80

10 Information value case: Life-saving diagnosis 81
- 10.1 Decision strategy . . . 81
- 10.2 Diagnosis . . . 82
- 10.3 Assessments imputed to D's diagnosis . . . 83
  - 10.3.1 Sensitivity analysis . . . 84
- 10.4 Formal underpinning diagnosis . . . 84
- 10.5 Does diagnosis warrant testing? . . . 85
- 10.6 Credit to ADT? . . . 86
- 10.7 Assignments . . . 87

10.8 Postscript 87

11 Expanded view: Should we teach decision making in school? 89
11.1 Reading, writing and decision making 89
11.2 Exercises 93
    11.2.1 Initial assignment 93
    11.2.2 Exercises for end of course 93

12 Epilogue — What next?   **95**
12.1 Making use of what you have learned 95
    12.1.1 Basic ADT logic 95
12.2 Developing your reasoning further 96

Bibliography   **99**

Index   **100**

Editors' preface

When Rex Brown died from the effects of pancreatic cancer in 2017, he had completed a near-final draft of this book, and had succeeded in obtaining a commitment from the publisher. He kept working on revisions until two days before his last breath. But many loose ends remained. He had wanted to add some additional chapters, but they were deemed unnecessary (including by Rex). There were several inconsistent references from chapter to chapter, and occasional missing examples.

Fortunately, the editor in charge at Taylor & Francis, Ceri McLardy, took the commitment to publish seriously. It was already clear that Jonathan (Jon) Baron, a collaborator of Rex's and himself an academic decision-scientist, was in the best position to do the substantive editing. Jon had begun this work before Rex died, but various other commitments got in the way (as well as Rex's obsessive rewriting).

Jon was a colleague and friend of Rex's for many years. He is happy to be able to help bring this, his last work, to fruition. He too is a true believer in the value of some sort of decision analysis for ordinary people, and his views on this were shaped by Rex, over many years.

Rex's daughter Karen, a journalist, helped work on editorial clarity, in part because she could recognize the times when what Rex meant did not exactly line up with what he wrote. She promised him, in his final months, that she would clean up the wording that his overtaxed brain and body didn't always get right, and that she would usher this project – the main focus of his final years, right behind his beloved family – to publication. The irony is, when Karen was growing up, Rex always wanted to explain to her the inner workings of applied decision theory, and she was usually too impatient to listen. Now she wishes she could ask more questions.

We have kept almost all of the substance of the original, except when it was redundant, while also making decisions about whether to keep or delete optional parts of the text. In a few cases, where we thought that Rex was over-stating something, we moderated the language, and we replaced a few references and examples with others that we thought were more current.

Finally, we thank the following for their generous editorial and logistical support (in alphabetical order): Leora Brown, Tamara Brown, Dalia Brown, Gregory Burnside, Marianne Chindgren, Michele Daly, Kobe Fox, Kathy Laskey, Brian Mancuso, Lucy Norton, Sam Norton, Sean Norton, and Sarah Shapiro.

Chapter 1

Introduction

As author, my goal is to offer readers, whatever their aptitude, interest or background, the skills to master a basic set of non-mathematical decision-making tools. Those skills will enable them to make wiser everyday decisions, both large and small, which will lead them to more satisfying lives. This text is a first step along that path, with a narrower scope: choices among options for personal dilemmas, such as whether to get married or whom to vote for. And unlike traditional decision-analysis, this book aims to bypass the burdensome technical methods that often turn off the everyday decider.

It is designed for use in a college course, but can also serve as "self-help" for the general reader.

1.1 Potential need

The prevalence of poor decision-making has been well-established and documented in best-sellers such as Daniel Kahneman's "Thinking, fast and slow" (2011). People make terrible decisions when they should have known better, and it may cost them dearly. They marry partners who are likely to make their lives miserable. They vote for a candidate even when they have clear evidence to the contrary that another candidate would govern more to their liking. There is plentiful research on how people make decisions already, but little on how people could make wiser decisions.

In the latter part of the 20th century, "Applied Decision Theory (ADT)" was developed, and the method was a substantial vogue in academic and professional circles. It was widely considered a universal key to wise decision making and widely adopted by business and major institutions. The idea was to analyze decisions in terms of options, the possible outcomes of choosing each option, their probabilities and a numerical measure of how much you value each outcome, its "utility." In this way, you could calculate an "expected utility" for each option and then choose the option that would potentially have the most value for you.

And yet, by the end of the century, ADT had lost its shine, its usefulness coming under fire. ADT was no longer the wave of the future. As with most theories, the devil was in the details. I did not share this blanket dismissal; my reading was that the ADT logic was sound and that this was not what was being challenged. Wise decisions conformed to decision theory norms, but the most useful way to make a decision did not necessarily take the form of any theoretical model. Other approaches such as intuition and feedback from real-world decision-making practice might bring the decider closer to the ADT ideal. Ordinary people could not use the tool, as then developed, cost-effectively. It was difficult to understand, laborious to implement, and rarely outperformed common sense. Operational methods needed to be adapted to fit cognitive capacity and practical requirements. The tool and its application needed more than logic; it needed psychology and feedback from practice. The methodology and its practitioners needed threefold skills in logic, psychology and practice feedback, rarely found in the decision making profession.

A highly regarded ADT-based text book by Hammond, Keeney and Raiffa (1999) features different methodological devices and topics and its contribution is complimentary, not competing with this book. It expounds difficult ideas with rare clarity. These authors and I share the same decision theory perspective. All four of us were part of the Harvard team that originally developed ADT in the 1960s. However, a distinguishing feature of the present book is to enhance (rather than replace) the usual decider's decision processes. Together, they could support a single course on prescriptive decision analysis.

1.2 Distinctive features of this text

- Qualitative counterparts to quantitative ADT models.

- Uses the totality of the decider (D's) knowledge rather than just what is called for by a single ADT analysis.

- Exercises students' real judgments (as opposed to hypothetical or other people's judgments used in traditional case studies).

- Illustrates argument with real examples.

- Combines multiple evaluations of the same judgment ("hybrid judgment").

- Provides real-world illustrations of decision methods.

- Introduces innovative analytical concepts (such as "ideal" judgment).

This text can be primary support for a short course in decision-making, as part of a variety of programs, such as psychology, management and philosophy. It can complement other decision-aiding texts, notably Hammond et al. (1999), which shares the same perspective, or "descriptive" behavioral decision texts, such as Baron's "Thinking and deciding" (2008).

Chapters are largely modular, and lend themselves to being taught as self-contained segments. Segments could be selective in emphasis: qualitative with or without quantitative treatment; personal vs. civic domains; factual vs. value judgments; case-studies vs. method exposition; more vs. less advanced material.

1.3 Vocabulary

The vocabulary here represents a significant modification to prevailing professional practice. I have found that students, clients and most others (including academics) who have not been indoctrinated in current practice are commonly confused or misled by much decision science language in common use. So, I have substituted, for current language, I hope, clearer language, which I have found communicates better. For example, I use "applied decision theory" in place of "decision analysis," which misleadingly suggests a generic way of analyzing a decision.

Language has proven to be such an impediment to communication as to put people off from learning or using modern decision aids. I have painful memories, from my early consulting days, of presenting what I thought was the clear analysis of an investment option to a company president. After listening for ten minutes, he muttered "gobbledygook!" and stormed out of the room and had nothing further to do with our work. My goal is to be welcoming and inclusive with the language, and certainly to keep everyone in the room.

1.4 Historical background

Before the 20th century, major scientific advances oriented toward decision-making had tended to be descriptive more than prescriptive, focusing on how the world works rather than how to make it work better. Decision circumstances changed slowly from year to year, so that decision practice could take its time to improve by trial and error. By the beginning of the 20th century, however, technology and other fields had begun to change rapidly. The life-and-death perils of poor decisions in World War II spurred the development of the quantitative decision tools of Operations Research. They were special-purpose tools (e.g., for locating enemy submarines) that may well have been decisive in winning the war.

After the War, Operations Research was adapted to industry, with some success for certain situations. These tended to be where options were complex and consequences were clearly defined, and involved processes that could be mathematically modeled (such as in production scheduling and transportation logistics). Progress in applying quantitative methods to choices involving a few clear-cut options with messy outcomes was a good deal slower. Analysis here competed less effectively with unaided humans, and deciders often did better by backing their own judgment.

The mid 20th century saw the development of general-purpose statistical decision theory, which can readily adapt to changing circumstances and, *in principle*, analyze any choice whatsoever. It does so by quantifying a decider's judgments about goals, options and outcomes, however ill-defined, and by inferring the preferred choice. Its practical application is *Applied Decision Theory (ADT)*.

In the early 1960s, research groups at Harvard and Stanford developed and promoted ADT as a universal methodology for improving rationality in a world where poor decisions were damaging lives and communities. Leading corporations (like Dupont, General Electric and Kodak), and then government departments (like Defense and Energy) began to apply decision analysis to their most challenging (and controversial) decisions. Many impressive successes were reported, and those of us who had been in decision analysis "on the ground floor" viewed decision analysis with a missionary zeal.

Decision analysis has passed through several overlapping phases, characterized by distinctive modes of aiding, each building on the earlier ones.

- The *theory* phase laid the foundations of the parent discipline, statistical decision theory (about 1950 to 1970)

- The *technique* phase focused on specific modeling procedures and sought illustrative applications (about 1960 to 1980)

- The *problem* phase selected from among available decision analytic techniques and adapted them to a particular class of problem, such as capital budgeting or environmental protection (about 1975 to 1990)

- The *use-and-user* phase addresses all requirements of useful aid in a given context, in which the focus is on usefulness to a particular decider and context.

Actual ADT practice by no means fits neatly into these categories. Their edges and timings are much more blurred, but they may give some insight into ADT's evolution. In some ways, I view this book as a manifesto for a "use-oriented" decision analysis revolution (rather than technique- or problem-oriented ADT). It draws equally on logic, people and practice skills (rather than on logic alone). I believe this development of emphasis can counter the pressures that have slowed down successful ADT practice in the past. ADT methodology, though useful as it stands, is still a work-in-progress.

1.4.1 ADT methods

At the time ADT was created, it was widely believed, in professional and academic circles, that any decider could improve his/her decision performance by acting on the implications of such a model. ADT modeling began to be taught to deciders-in-training throughout professional education (Brown, Kahr & Peterson, 1974). It was expected that it was only a matter of time before ADT would become standard practice in management and other applied domains. Indeed many of the nation's major business and government or organizations began incorporating ADT into their decisions (Brown & Ulvila, 1982).

However, ADT, as originally and still generally practiced, has since lost its once-booming interest. Indeed, it is no longer a required course at Harvard Business School, where it originated. It has proven of little practical value, due to three fatal flaws: failure to adapt to human cognitive capacity; apparent cost in time of ADT analyses; and disregard of knowledge not called for by an ADT model. The action that ADT analysis favors is based on a single numerical model and largely ignores other sources of wisdom, such as intuition, others' advice, and alternative analyses. As a result, ADT analyses do not usually improve on unaided decisions — at least not enough to be worth the trouble.

The root cause of these technical ADT problems that impede useful prescription (as opposed to description) is that ADT tools have been developed, applied and taught by mathematically oriented statisticians and the like, who lack both relevant psychology training and familiarity with the real-world contexts. They can check for logical soundness, but not cognitive or organizational fit, or cost-beneficial balance. Academic career incentives do not motivate them to do otherwise.

1.4.2 My own ADT evolution

My first job out of college, in 1958, was in management consulting, which included trying to help clients make decisions in the 1960s. Seeking unsuccessfully some logical discipline to support my advice, I heard of relevant work on ADT led by statisticians Howard Raiffa and Robert Schlaifer (1961) at Harvard. I joined their group and spent five years absorbing their ADT methods, teaching these to MBAs.

1.4. HISTORICAL BACKGROUND

When I tried these out on business clients, I was disappointed (as others were) to find that, in spite of ADT being logically compelling real, deciders rarely used it or found it useful. I attributed this to the fact that our version of ADT did not take into account human capacities and limitations.

So I moved to the University of Michigan for the next four years, to learn from psychologists, led by Ward Edwards, who were describing how people make decisions and especially their logical flaws. I found that enhanced knowledge improved my decision-aiding ability somewhat. I incorporated it into a decision-aiding course in collaboration with an interested psychologist and statistician, who complemented my earlier practical decision-airing experience (Brown, et al., 1974). This variant of ADT was an improvement, but it still lacked explicit adaptation of decision tools to human capacities.

In 1973, I returned to my original consulting career. I spent the next two decades aiding professional policy-makers (such as business managers and government officials), developing new decision tools (often funded by research agencies), teaching future managers (such as MBAs) and educating the general public (through the press and broadcasts). I interleaved consulting to practicing deciders and serving on relevant faculties, notably statistical decision theory (University College, London), behavioral decision theory (London School of Economics) and management (Harvard, Michigan, George Mason) and, where appropriate, collaborating with others within them.

By the time I retired from paid employment in the mid-1990s, I had come to believe that the private decider (the "common man") stood to benefit from learning to make wiser decisions at least as much as the institutional decider. Accordingly, I had begun to re-direct my research, teaching and applied efforts from professional to private decisions. I set about adapting my professional experience to improving private decisions, by teaching the young (and the adults they would become) how to make wiser decisions. This text encapsulates much of what I learned.

Psychologist Jon Baron and I developed teaching materials and pilot-taught them in local schools. (See Baron & Brown, 1991.) This experience confirmed that private deciders needed aid as much as professional deciders, but in a substantially different form. For example, private deciders do not generally have to quantify thesir reasoning explicitly, nor validate their reasoning to others, which is a key feature of professional decision aiding.

1.5 Looking ahead

Digesting the decision principles and tools presented won't guarantee avoiding the pitfalls that, unaided, intuition and informal reasoning might cause. But it can move students perceptibly in the right direction, and do at least as much good as spending the same effort on most other academic subjects would. It may also give them a general clarity of thought that enhances what they get from other courses (like understanding Hamlet's "To be or not to be" dilemma, in English Literature; or the workings of natural selection in Evolutionary Biology; or the pros and cons of nuclear energy and fossil fuel in Environmental Studies). Even when this training does not improve much on good common sense, it should clarify the underlying logic, make deciders more confident in their conclusions (when confidence is warranted), and help explain their reasoning convincingly to others.

Chapter 2

Decision making

In this chapter, I discuss the objectives that I hope this text will meet for you as a private decider.

2.1 What is a decision?

When you face a *dilemma*, that is, a tricky situation that may call for action, you normally proceed in two steps. You make a decision, which is an *intent* to act (or not), and you may, or may not, act on that intent. For example, ♦[1] you "decide" to diet, but don't manage to.

The simplest decision is a choice among a few clearly identified options, and that is the main focus here. However, the same principles apply to the common case where there are numerous options, not all of which are immediately apparent.

Most decisions are made informally, whether intuitively without deliberate thought, or based on careful reflection. However, over the centuries people have recognized the need for better and more defensible decisions and have tried to develop systematic and structured aids. I will try to advance that enterprise.

---

[1] A diamond indicates an example, throughout this book.

2.1.1 Ideal decision

If you were *all-wise*, you would make choices that maximize the prospect of your future *satisfaction*. Satisfaction is an elusive but essential concept, variously referred to as happiness, welfare, desirability and, when quantified, *utility*. For our purposes, it is whatever serves your *preferences* or values.

You would have achieved *ideal* wisdom when you have perfectly processed your entire *mind-content*, i.e., absolutely everything that you know and feel. You would make the same decision as you *would have* by following an impeccable decision process, regardless of how well you actually do decide. If you were such a person, you would not have much use for this book. Fortunately for those of us who make a living trying to improve decisions, such all-wise deciders are rare indeed.

Ideal wisdom is not remotely achievable, given your normal cognitive limitations and the current state of the art of logical analysis. We are all, to some extent, unwise. However, ideal wisdom serves as a distant beacon toward which we can strive. *Rationality* is the achievement of wisdom through some deliberate reasoning process, such as I will propose here.

♦ Reg is a low-wage worker deciding whether to contribute money to public radio. If he asks himself "what's in it for me?" he concludes "not enough" and decides to saves his money. (He happens not to feel guilty about not "doing his duty" to contribute, even though he listens to public radio.) Yet he "irrationally" does contribute, because he is "that kind of person", i.e., programmed that way, even though it is against his immediate interests (saving money).

This is a perverse example of a "cognitive flaw." After all, many of us consider contributing to public radio to be a decision made largely for the public good; the perversity is that, based on Reg's own values, which include saving as much money as possible, he would act against that perceived good.

2.1.2 Value of wisdom

Much of the time, you feel comfortable about the option you end up choosing in a given dilemma. However, you sometimes worry that you could have better integrated what you know and what you want, and thus given yourself a better chance of meeting your goals. And when you know what to do, you may still have trouble communicating your reasoning to others whose perspective differs.

Muddled thinking — or at least the lack of responsible reflection — is widespread and may cause serious, but avoidable, harm.

♦ You may be led to support legislation promoted by special interests (trade restriction? tax relief?).

♦ You support (or oppose) health regulations without responsibly trading off risks and costs.

♦ You take the first appealing job that comes along. You miss out on a better option and deprive a more appropriate employer of your talents.

2.2 Requirements of wise action

To act wisely consistently, you need to be good at several things. You must:

- Recognize when some action may be called for.

- Identify promising options.

- If necessary, seek information on those options.

- Choose the best option.

- Act on that choice.

- Follow through effectively.

This book is aimed mainly at helping you with the fourth task – choosing among options — with some attention to the third task — knowing when and what information to seek first. The other tasks are just as important, but I am not addressing them much here. (Reading this and doing the exercises may confer benefits beyond helping you to act wisely, such as improving your critical thinking in general.)

2.2.1 Accounting for all available knowledge

A key feature of a wise choice is that it draws on all your knowledge, not just what happens to be used in any particular analysis. The fact that an analysis is complex and logically coherent may not produce a choice that is wiser than unaided judgment, if the latter properly draws on more knowledge — that is, draws on your entire mind-content.

♦ Before you decide how much to pay for a house, don't just look at the cost per square foot; also consult your realtor and check how much similar houses have been selling for.

♦ Is it not unknown for a meteorologist to announce only a 70% chance of rain, while outside his window rain drenches the region.

♦ Ford Motor Company UK felt they had too many car parts depots in the London area. A prestigious university Operations Research group developed a state-of-the-art math model that indicated that only four of Ford's existing seven depots were needed. Accordingly, Ford closed three depots, with disastrous results. The four remaining depots proved completely inadequate for the demand. It turned out that depot capacity had been grossly overestimated. The analysts had casually calculated usable depot capacity as height times length times width, as if the depot were a box that could be filled to the top. They did not take inevitable dead space into account, though they would have if they had taken time to think seriously about it, or checked with a depot manager. As a result, predicting the outcomes of closing down depots was grossly distorted. The analysts were high-powered statisticians, with little knowledge of (or, I suspect, interest in) the nitty gritty of inventory management. They presumably did not see the need to take all relevant information into account, in order to reach a wise decision.

2.2.2 Wisdom is no substitute for knowledge

A wise decision depends both on how good the knowledge you use is, and on how well you reason from that knowledge. Here, we focus mainly on the second, though it is important to recognize that the first is often more important.

♦ If you are deciding whether to take on an unfamiliar room-mate, it may be more useful to seek more information about her background than to ponder over the little you already know about her.

♦ People often mistakenly assume that knowing something about rationality puts me in a position to instantly give good advice on what to do, when I have only the sketchiest knowledge of the problem. "Garbage in, garbage out" — meaning, using poor information leads to poor decisions — applies to the most rational people, except that they ought to know better than to say anything at all.

♦ *Civic example:* I was on a radio talk show and a listener called in to ask me if I thought the US should invade Afghanistan. I explained that unless I had carefully analyzed and researched the problem, my opinion would be worth no more than the listener's, and probably less.

2.2.3 Good outcomes may not imply wise decisions

An unsatisfactory outcome does not necessarily mean your decision was unwise — or vice versa. It may be wise to wear a seatbelt, but, against the odds, a seatbelt could kill you (say, by trapping you under water). Nevertheless, if the outcomes of my decisions are consistently worse than yours, it is a useful, if inconclusive, indication that you are wiser than me.

♦ *Bachelor Joe example:*[2] Bachelor Joe is agonizing over whether to marry Jane or Lulu. (For the sake of argument, let's just assume the two women have given him that choice in a complex love triangle.) In the end, he marries Lulu, whom he finds more attractive, contrary to convincing indications that, by her character, she would make a terrible mate for Joe — and so she proves to be. However, they produce two wonderful children, who make Joe's life happier than he could have expected with Jane, in spite of the marital strife that he endures with Lulu, as mutual acquaintances had predicted. By hindsight, he does not regret his decision to marry Lulu. He acknowledges later that it was an unwise decision, in the light of evidence available to him at the time. He just got lucky.

♦ After Jane is left by Joe, she falls hopelessly in love with Jasper, a charming, recently divorced man, who is admired by all her friends. They have a whirlwind romance and three months after meeting, in which they discover many shared interests, she marries him. He turns out to be an abuser who makes her life a misery. She may not have acted unwisely in marrying him. She may have made best use of the misleading information she had, if her only options were to marry him or not. She just got unlucky.

---

[2]This hypothetical example will reappear throughout this text, often with different assumptions.

2.3 Current decision practice

2.3.1 The world of private decisions

We make decisions all the time and they shape the quality of our lives, but not always for the better. Most day-to-day decisions, such as what to have for dinner, are too small to do more than act on intuition, without aid. However, cumulatively, even these small decisions have an important impact on our lives. Decision-making is an immense, complex and life-long undertaking. Improving it, even modestly, is surely as important as, say, learning to drive or knowing geography.

Over the centuries, people have recognized the need for better and more defensible decisions and have sought systematic and structured aid. Benjamin Franklin, for example, suggested a useful way of canceling out equivalent pros and cons of options, after making lists of each, until the best option becomes clear, because what is left is mostly pros, or cons. Such sporadic approaches foreshadowed intensive efforts, a century later, to develop various forms of prescriptive decision analysis (PDA), including *applied decision theory (ADT)*[3], the logical core of the approach presented here. The practical use of ADT, by the late 20th century, had penetrated virtually all segments of professional decision-making (especially business, government and medicine), so far with mixed results. However, to date, its application to private decisions has been modest (Brown, 2012).

2.3.2 Room for improvement

It is no easy thing to improve on how humans have been making personal decisions for thousands of years. Natural selection must have winnowed out the least effective of our forebears.

♦ The wise caveman who, when faced with a mortal enemy, makes the right call between fight and flight may live to reproduce another day, outliving his less wise neighbor.

---

[3] Also known as "decision analysis", which misleadingly suggests something much broader.

2.3. CURRENT DECISION PRACTICE

Ideal judgment

There is general agreement that decision practice throughout society still has plenty of room for improvement. We can all call to mind examples of people making ill-considered decisions where "they should have known better" — i.e., they had knowledge or sentiments that logically implied some *ideal judgment*. Your evaluation of how far your actual judgment falls short of your ideal is an indication of how wise you think your judgment is and how much room for improving your judgment you can imagine. (See Figure 1.)

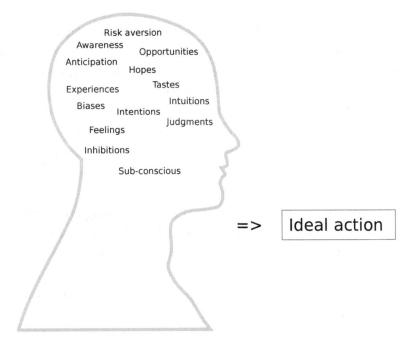

Figure 1: Ideal action inherent in your current mind contents.

2.3.3 Motivation for wisdom

Given that unwise action is widespread, what harm does it do? What benefit can you expect to gain by acting more rationally? Is it worth learning analytic tools that will make you more rational? I know of no empirical research that has produced definitive answers. All I can say is anecdotal and impressionistic. I have seen many cases where I am persuaded that someone made an unwise decision and had a bad outcome that I believe (though cannot prove) would have been averted had they made a wiser choice.

However much acting wisely may be in your self-interest, I have found it most difficult to persuade people to take it seriously.

♦ In political campaigns, voters seem more interested in who *is* winning than who they *want* to win. Broadcasters devote much more air time to tracking the fortunes of contending policies than to evaluating their pros and cons. Audiences would sooner watch a fight than listen to a lecture. And political science has shown us that horse-race polling can actually become a self-fulfilling prophecy.

♦ *A personal testimonial:* Using the principles to be presented here may have added several years to my life (Chapter 10). Spring 2011, I told my doctor, a highly regarded primary care physician, that I was concerned that my weight had been dropping with ominous regularity for the past several months. He told me that was nothing unusual, especially as I had been observing a stricter diet, and he recommended "watchful waiting". Not completely reassured, I elicited his medical expertise along with my own value judgments and more nuanced reading of my symptoms and background, according to a rigorous decision logic.

I inferred that my chance of having a life-threatening condition was high enough to justify major diagnostic testing. The doctor was persuaded and he ordered the tests. These indicated I had stage 2 pancreatic cancer, just in time to perform successful surgery. Six years later, I was still completely cancer-free...: ("Watchful waiting" would almost certainly have been too late to save my life for this period.)

2.3.4 Opposition to rationality

Advocates of certain courses of action (such as advertisers and politicians) to be taken by others (such as consumers and voters) are motivated to discourage any thinking that might lead them to contrary action, including rational decision-making.

♦ Long ago, the Rector of Leningrad University (in the old Soviet days) told me they tried teaching a variant of rational choice, but students kept coming to the "wrong" conclusions (that is, conclusions that the Rectors did not approve), so the instruction was dropped.

2.3. CURRENT DECISION PRACTICE

♦ I have gotten similar responses from religious parents who feared that decision analysis would lead the children to conclusions that contradicted religious teaching.

♦ A student in the Environmental Science department of a major university told me his professors discouraged study of "rational choice". Presumably, they considered environmental protection to be a sacrosanct cause, not to be balanced against any other considerations (i.e., economic).

♦ More defensibly, you might motivate a friend to act on a rationally defensible New Year's resolution (say, to exercise), using an "irrationally" strong argument, i.e., by exaggerating the case (say, by overstating the health benefits of exercise). You believe that evaluating pros and cons realistically would indicate the same decision, but you can't trust your friend to act on it. So you feel justified in this well-intentioned, but deceptive, advocacy.

2.3.5 Raw material: The contents of your mind

The wisdom of your reasoning is limited by the contents of your mind and what you can learn to add to those contents. Your *mind-contents* may be quite messy, and include thoughts, feelings, anticipations, recollections, misgivings and sub-conscious activity — often quite jumbled — which may change at any time. (See Figure 1.)

I hope to help you to extract wise decisions from your mind-contents, typically building on your familiar existing ways of thinking.

2.3.6 Decision vs. action

This text is about improving decisions and the judgment and reasoning on which they are based. However, faulty decision-making is not the only reason we act unwisely. We can be stubborn, intimidated or paralyzed. I don't know how to combat such human frailty (as my own failure to diet and exercise testifies). Nevertheless, *thinking* smart is still important enough to be worth improving. This is what I will focus on — how to make up your mind, regardless of how you act on it.

Deciding *wisely* and acting effectively means you never have to "kick yourself" for having been unwise. You do the best you can with what you know and feel. You can make the right decisions, but that doesn't guarantee you will act on them.

♦ A farmer was asked on TV why he wasn't taking a local course on agricultural practice. He replied, "I'm not farming now as well as I know how." Again, knowing what to do is no guarantee of doing it.

Head, heart or backbone may fail you. Here we are concerned primarily about getting the head part right. Wiser decisions may foster wiser action, but they do not guarantee it. Thinking smart is no guarantee of acting smart, but it is still important enough to be worth improving. How much time and effort to spend on improving your thinking vs. improving your acting on it it is itself a decision that could benefit from this text.

2.3.7 Precept vs. process

Processes are not the same as *precepts*, in the sense of telling anyone what they should actually do.

♦ When I first approached NICHHD (National Institute of Child Health and Human Development) for a grant to develop adolescent decision skills, they assumed it was to help young people make "correct" choices: such as, don't do drugs or sex, but do stay in school. However, NICHHD went along (possibly with some regrets) with my proposal to teach people a process for achieving their own goals.

To enhance the *process* from which a good decision may emerge, you need to specify what questions to ask and what to do with the answers. It will not necessarily lead to the outcomes you prefer. But you can hope a coherent thinking process will give you the best chance to obtain a good outcome. You are not receiving a fish, but learning how to fish.

2.4 Decider roles

You make decisions — or will in the future — in different capacities. They may be professional (say working for an organization) or private (on your own behalf). We are concerned only with *private* decisions, which, in turn, can be either *personal*, that is, in your own interest (say, choosing a mate) or civic, that is, in the interest of a community (such as voting for president).

2.4. DECIDER ROLES

2.4.1 Personal decisions

Most daily personal decisions are straightforward. Most are small in scope (such as whether to go on a date), but taken together they determine much of the quality of your life. Most only require you to make up your own mind on the basis of your personal interests and what you know, and take no more than a few hours of reflection. However, with some decisions, a misstep (like a bad marriage) may be devastating and merit greater effort.

Personal choices include:

- Should you train to be a dancer? (You enjoy dancing, but it has a low chance of economic stability.)

- Should you invest in stocks, bonds or real estate? (These differ in high vs. low risk, compared with high vs. low returns.)

- Should you put your Down's syndrome child in an institution or care for him at home? (What is best for the child may not be best for the rest of the family.)

- Should you have an arthritic hip replaced now or wait a few years? (You need to weigh cost and inconvenience against the chance of surgical success.)

Some types of personal decisions, though individually small, are relevant to so many people that they justify major public aiding efforts (such as career advisory services).

2.4.2 Civic decisions

Civic choices, like personal choices, are private, in that they are yours alone to make. However, you make them as a citizen on behalf of a larger community, such as the electorate, whose interests you serve. You vote for a political candidate or for a public policy position, say, on healthcare.

As a citizen, you have no direct responsibility for any communal action.

♦ You choose whether you are "pro-choice" or "pro-life" on abortion; whether you favor military action abroad; whether nuclear power should be stopped; whether your country should pressure Israel to leave the West Bank. But your impact on what gets done is minimal.

Civic decisions differ from personal decisions in a number of ways. They are much less important in your life than personal decisions. However, the interest of the community as a whole (including yourself) will be better if all members make sound civic decisions. Getting citizens to vote for policies and politicians that best serve the aggregate interests of all those affected would certainly benefit society immensely.

However, paradoxically, training citizens to act more rationally may actually be a disservice to society. Realistically, any individual's vote will have minimal impact on who gets elected or what legislation is passed. A rational voter may give higher priority to "feeling good" or to the symbolism of supporting a particular ideology than impotently casting his one vote in the best public interest.

♦ In the presidential election of 2000, many liberal environmentalists voted for their favorite, Ralph Nader, whom they acknowledged would not win. Their desire to "vote their conscience" thus helped assure the victory of President Bush over their second choice, Al Gore, whom they would have much preferred.

2.4.3 Civic vs. professional decisions

A civic decider, such as a voter, has only remote influence over what action is taken, and can rarely spend more than an hour or two thinking about it. He is normally under no particular pressure to take great care weighing the pros and cons of a civic choice and would only use the simplest of decision tools, if any. He has the luxury of ill-considered (but perhaps emotionally satisfying) positions, at little personal cost; but possibly, if there are many like-minded voters, with great social impact.

On the other hand, a professional decider is acting for someone else (such as society, an employer or a client), say, as a manager, military commander, public servant or physician. He is directly responsible for such choices and may spend several months analyzing them before making a commitment.

♦ A congressional staffer (a professional decider) may spend weeks evaluating legislation, and call upon extensive specialized analysis and information.

On the face of it, professional decisions look much like civic decisions, in that they can both refer to the same public action. However, the stakes are normally so much higher for professional decisions (in part, because the power is greater with the professional). Therefore, the social value of aiding a professional decision is greater than for private decisions, because the cost of error and the value of reducing it are greater.

In contrast to private decision-making, the practice and teaching of professional decision-aiding is already well-advanced, especially in business, public policy and medicine, and teaching materials.

Although this text does not address professional decisions directly, I believe learning first how to make wiser private (including personal) decisions is your most effective preparation for learning professional decisions. However, these have additional levels of complexity, and decision practice differs from one profession to another.

2.5 Student end product

From this text you should develop some immediately useful mental skills. You should be better able to:

- Make rational choices *informally* and appreciate the considerations that go into them, building on how an intelligent person, experienced in making good decisions, already thinks.[4]

- Build simple, logically sound models. These models need not be particularly ambitious technically, compared with what you might find, say, in a typical Operations Research course. They help illuminate your informal thinking, but do not supersede it.

- Integrate such models effectively into informal decision processes, so that they yield better and more useful choices. (It takes an unusually good student to do this successfully.)

- Effectively specify and use more authoritative models (to be developed and applied by decision analysis specialists), to use on your own decisions, professional or private.

---

[4]Nobel laureate Richard Feynman, having performed countless complex mathematical computations over the years, got to the point where he could *intuit* the answer to within a few percentages of the correct answer, without doing any calculations. Enough practice with formal decision analyses may do the same for our intuitive decision making.

This text alternates technical material with real (or at least realistic) case studies. These include a term project where you will analyze a live civic choice, of your choosing, drawing on decision-aiding methods as you learn them as (and if) appropriate.

2.6 Exercises

2.6.1 Bad outcome — good decision

1. Dolly complains that husband Reg makes poor investment decisions. For example, they had stocks that Reg failed to sell as its value dropped disastrously over several months. Periodically, she would say "you should have sold last week; of course, it's too late now." Is Dolly's criticism valid?

2. A UK prime minister dissolved parliament and called for general elections before he constitutionally had to. He loses the election and realizes that he would probably have done well to wait. Was it an unwise choice?

3. The US government is spending billions of dollars to find a technically and politically acceptable US site to dispose of high-level nuclear waste. So far, the effort has been without much prospect of success. It has been proposed that the US government offer nuclear waste to Ethiopia for enough money to eliminate starvation in Ethiopia in the foreseeable future. Under what circumstances might this be a responsible and rational decision for the US government?

4. Describe briefly a real past choice among clear-cut options, made by either you or someone you know, of each of the following types:

   a. The choice was unwise, and turned out badly.

   b. The choice was unwise, but it turned out well.

   c. The choice was wise, but it turned out badly.

Why was the choice wise or not in each case?

5. Pole-vaulter Gold medal hopeful Robert O'Brien opted to delay his first jump in 1992 Olympic trials in Barcelona, until the bar had been raised dangerously high. He wanted to preserve his energy. However, he failed at his first try and so did not qualify. Was his delay decision unwise, or simply unlucky by hindsight?

2.6.2 Decider role

Which of the following are professional, personal and/or civic decisions?

a. Accountant D is making up his mind on which candidate to support in the next presidential election.

b. Supreme Court Justice D is deciding whether present election laws are constitutional.

c. D is deciding whether to accept a job offer in Timbuktu.

Chapter 3

Term project: Evaluating a policy proposal

3.1 Task

Singly or in groups, you are to work on a civic proposal, chapter-by-chapter, topic-by-topic, throughout the rest of this text. You apply techniques and concepts as you learn them, culminating in a choice that you really trust, and would be prepared to act on. You need not limit your reasoning to course material, except to the extent that it helps. Blind faith in what pops out of a structured analysis is certainly not what I want to impart. Integrating what you learn into your everyday decisions is.

3.2 Nature of issues to be evaluated

Propose the wording of a binding referendum on which you will vote. It is important to pick a proposal that is not trivially obvious to most students, in order to foster meaningful deliberation accordingly, an initial vote should show some balance between those inclined for and against. If one of the options is considered "politically incorrect", students may be embarrassed to support it openly, in a show of hands. This would favor anonymous voting (as in a real election). You should only address issues bearing on what public action you would like to see taken; not, for example, what to do about lobbying or other political impediments.

3.3 Project activity during course

You will develop the project technically, much as you would if you were really going to vote and as if it were a real effort to aid a decision, with two significant exceptions.

First you must limit yourself to using factual knowledge you already have. If you were really facing the problem in the real world, acquiring new knowledge before deciding might be sensible, but for learning PDA (prescriptive decision analysis), it would not be the best use of your time. If you are completely unfamiliar with the issue, talking to, at most, one outside person familiar with the issue, or reading some brief press discussion, should provide enough background. The emphasis should be on analyzing what you already have in your head.[1]

Second, you will spend a good deal more time working on this project than you would normally want to spend on a vote. However, you are developing skills that can be transferred later to professional decisions which may call for much more effort or time. (The lower level of effort called for on this project may be appropriate for a *preliminary* analysis by the professional decider you may become, to be followed by more intensive study by a technical specialist.)

3.4 Sample proposals

1. Government will not pay ransom for freedom of hostages.

2. Voting in national elections is to be mandatory (as in Australia).

3. How legislators are elected should not be controlled nor influenced by the legislature, but by an independent body.

4. Authorize major research to develop a pharmaceutical which makes

---

[1] You may, in fact, develop more knowledge about the issue than you started with, as you work on the project. This will result in improved choice that is *not* due to our tools — except to the extent that tools alert you to what new knowledge is needed.

people feel happy without the bad side effects (that banned drugs produce).

5. Tampering with or withholding evidence in a criminal trial (i.e., by police or prosecuting attorneys) will be treated as a criminal offense.

6. Alcoholic beverages are prohibited on any university-affiliated premises.

7. Retirement age at which citizens can draw a government pension is to be raised by two years.

8. Gun ownership is subject to a license, with requirements comparable to a driver's license.

9. The age of sexual "consent" is lowered from 18 to 16.

10. The US president is to be paid a salary comparable to the president of a major corporation.

3.5 Final report

The object of a final report will be to demonstrate your ability to apply analytic tools covered in this course to a real problem. (The technical mastery of the tools is tested elsewhere.) The report should be about ten pages, including figures.

3.5.1 Suggested report sections

1. Task statement, including: decision context (such as who the decider is).

2. Initial informal evaluation including preliminary choice (see initial assignment below).

3. Formal analysis including: analytic strategy, summary of steps and results.

4. Adjust conclusions to account for any considerations not previously addressed.

5. Statement of final decision.

6. Reference to any relevant work you have submitted earlier.

3.6 Standing assignment after each chapter

Apply what you have learned in each chapter, at the time it is assigned, to this term project.

To get you started, here are two things you can do now:

1. Would you vote YES or NO, on your policy proposal if you had to make an immediate choice, based solely on your informal reflection? Summarize your reasons in three lines or less.

2. List up to three factual outcome uncertainties and up to three value dimensions you would consider in casting your vote on the term project issue. Which of these six (or fewer) considerations would weigh most heavily in your choice and deserve most of your attention?

Chapter 4

Qualitative decision aids

This chapter overviews qualitative approaches to wise decision aiding. Decision aid of some kind has been around at least since the Delphic oracles "helped" ancient Greeks to time their sacrificial choices. Although intuition and informal reflection have been, and no doubt always will be, the dominant modes of private decision making, many ways to improve on them are available, both common-place and novel.

4.1 Types of decision aid

Decision aids come in a large hierarchy of forms, variants and levels of complexity.

4.1.1 Prescriptive decision analysis

Prescriptive decision analysis (PDA) is any systematic reasoning, designed to improve your decision-making. It encompasses decision aids as PDA tools, which identify preferred action. It prescribes what you *should* do rather than *describe* what you *actually* do. A medical analogy: biology is descriptive; medicine is prescriptive. PDA covers both *primary* decisions (e.g., you buy a house) and *informational* decisions (you first check out the real-estate market).

Your preferred decision approach is the one you judge will come closest to some imaginary *ideal* analysis of your mind-content, such as an *all-wise* decider could produce, subject to cost. The object of PDA is to extract as much relevant material as possible from the content of your mind, to this end. You may use more than one decision approach. You can think of it as a way of getting a "second opinion" about a decision you have already tentatively made. But the second opinion comes from someone with the same beliefs and values as the first: you.

PDA is a broad class of aid and comes in a hierarchy of forms, variants and levels of complexity. A richly-stocked and well-established kit of structured prescriptive tools is available, suited to all levels of intellectual maturity and training. It includes methods that both do and do not include numbers.

PDA can be unstructured, such as "pattern recognition" or "lateral thinking". However, we are mainly concerned with structured forms, both qualitative and quantitative. Study of any practice-oriented area, like medicine or engineering, can address two kinds of question: how the world works (including its people), and what can be done about it. *Cognitive decision theory* is concerned with the first of these, i.e., how people *do* decide. In this book, on the other hand, we are concerned primarily with *prescription*, i.e., how people *should* decide (though this must also take into account how they *do* decide now).

You should learn *how* to decide, rather than be prompted to take any specific action.

♦ Joe may be guided by PDA on how to choose between Jane and Lulu, even if he does not use PDA for his final decision about which one to ask to marry him, or even if he decides that his initial use of PDA is too simplistic. His "second-order" prescription may then take into account both the PDA and other considerations that it had ignored.

A PDA is like a "second opinion" from a different physician, but both opinions come from the same person.

We are concerned with *prescribing methods* and general principles for decision making, such as: "Distinguish facts from values, and don't let your values affect your facts", or "When you are evaluating a course of action, ask yourself 'Compared to what?'"

Some PDA does not directly indicate your preferred action. It can consist of becoming aware of common psychological sources of unwise judgment and take them informally into account ("cognitive vigilance"). Descriptive PDA, such as cognitive heuristics and biases, are well-researched and reported), so I have only lightly covered them here (in the context of predicting the uncertain outcome of decision-aiding methods).

PDA also includes physical measures such as "sleeping on it", rest, exercise and taking nutritional supplements. For example, If you are to decide tomorrow morning whether to accept an urgent job offer, you may gain more by getting a good night's rest than by staying up late to analyze your options in detail.

4.1.2 Applied decision theory

This chapter focuses on qualitative principles of sound decision reasoning, which are based *implicitly* on the logic of *applied decision theory* (ADT). Explicit ADT involves quantifying your judgments and inferring the preferred action they imply (Chapter 5). Virtually any defensible decision aid can be expressed in quantitative ADT terms, though it may not be worth taking the trouble to do so.

4.2 Projection

A common form of prescriptive decision analysis (PDA) is *projection*, that is, anticipating, one way or another, what may happen if an option is acted upon and comparing its prospective satisfaction.

4.2.1 Visualizing option aftermaths

Visualizing option aftermaths is a projection decision aid which has you imagine the scenario resulting from an option. You visualize a plausible aftermath of each option and evaluate which you would prefer to experience.

♦ In the classic movie, "It's a Wonderful Life", the James Stewart character is deciding whether or not to commit suicide, on the grounds that the world would be a better place without him. A helpful Angel shows him the aftermaths of his two options: life with him and without him. He clearly prefers the former aftermath and decides against suicide.

Two types of judgment are required: how to visualize an aftermath scenario in meaningful terms, and how satisfactory it is.

Reference comparison (comparison of both options to the same reference point) may help you to make the latter value judgment, particularly if a comparable aftermath is within your range of experience.

♦ A decider contemplating a second marriage can ask herself, "Was I happier single or married to my first husband, and is there anything about my current choice that favors/disfavors remarriage?"

Imagine you are watching a movie of your life under each option you are considering, with all the detail and atmosphere that a movie would convey. At the end of such "viewings", judge which aftermath you would sooner have lived.

You may need to take into account your attitude toward risk. How *uncertain* are you about the option aftermaths, in terms of how much they appeal to you? Is your uncertainty about the aftermath of the seemingly preferred option so much greater that it would switch your choice, given your risk aversion?

♦ *Bachelor example.* Joe imagines, in vivid detail, the course of married life with Jane and Lulu. He summarizes them as follows (though a summary cannot capture all the specifics that really matter to him). The Lulu aftermath starts with a couple of years of interesting social life, satisfying sex and mutual fun. They have a couple of children and Lulu is an adequate but not particularly caring mother, nor particularly engaged in the children's lives, as they get older. Lulu proves to be a great social asset in Joe's work. However, their relations with each other become increasingly strained and they drift apart and eventually nastily divorce.

With Jane, life is initially rather boring, but she proves a reliable companion and great mother. She stands by him when his career falters. He has a brief extramarital affair, which the marriage survives. Joe's life continues dully, but not unhappily. Joe slightly prefers the Lulu aftermath. However, it is more uncertain than for Jane (whom he has known for years). He is afraid that the Lulu aftermath could well be much worse than his best guess, and Jane is a surer bet. He dislikes the risk and chooses Jane.

A problem with *visualizing aftermaths* as a decision aid is the difficulty of explaining the rationale for your choice to someone else. Your verbal description of an aftermath, even quite lengthy, cannot convey enough of the fine-grained texture of the experience for you (let alone for someone else) to judge how satisfying it would be.

In your imagination, you may be able to conjure up something like a movie of an aftermath. Describing the experience in words cannot really capture how happy you are with the experience, as well as an imagined movie.

♦ A brief publicity write-up of the movie, "Amadeus", about the last years of Mozart's life, cannot really tell you if you would prefer his life to yours, whereas viewing the movie itself might.

♦ "Boyhood" is a popular movie depicting the life of a young man from preteen to adult hood. Suppose my fairy godmother gave me the option of trading my youth for his. Would I take it? If all I had to go on was a one paragraph synopsis of the movie, I have absolutely no idea. However, after seeing the movie I am pretty certain I would stay with my own life, particularly since I can be much less certain about how happy I would be with his boyhood compared with mine.

4.3 Anchoring judgments

Another simple PDA step is to anchor your decision to some comparable, but more accessible, *reference* judgment (as just mentioned), which can be adjusted to take account of adjudged differences between it and the target case, due to any special circumstances.

4.3.1 Comparing this dilemma with past decisions

One variant is to compare the decision at hand with some similar choice in your or others' past experience, evaluate how well they turned out and adjust this evaluation for any perceived mismatch between that reference experience and your target option.

♦ *Bachelor example:* Joe's friend Marty married a wealthy woman much like Jane and the marriage worked out just fine. However, Joe thinks Marty is less secure than he is and would value Jane's wealth more highly than Joe does. So Joe discounts Marty's experience and chooses Lulu.

♦ *Cheat?* Reg is considering having an extramarital romance. He focuses his attention on what might happen if he is or isn't caught, bearing in mind guilt, sexual pleasure and marital upset. He remembers saying to himself, "I'll never go through this again", after reflecting on similar episodes in his past and decides to remain faithful to his wife.

4.3.2 "Expert" opinion

You also might seek guidance from a respected source of wisdom, such as friends, relatives and experts, and again allow for any special circumstances in this case.

♦ Joe's mother advises him to pursue Jane. He thinks his mother's values are more old-fashioned than his, but not enough to override her advice. So, he pursues Jane.

4.3.3 Adapted decision rules

You can also take into account established decision rules (if ... then ...) relating to the same class of decisions. But you need not follow such rules as if they were absolute. You adjust your conclusion for any relevant ways your target choice is unrepresentative of the general class.

♦ The rule "Never drink coffee after 6 PM" might yield when you have to stay up very late to get some work done.

♦ To get a mammogram, until recently, the US government guidelines advised women under 50 against having routine mammogram tests for breast cancer. In early 2014, in response to new research findings, this decision rule was made more specific (soundly, I suspect). Women between 40 and 50 were now advised to have mammograms under certain circumstances, to be jointly determined by patient and physician.

A variant of decision rules *is conventional wisdom*.

♦ Our beloved 18-year-old family cat, Scamper, was dying of a failed liver. My pre-teen daughters tearfully urged me to get her a liver transplant (which they reckoned we could afford financially). Out of respect for their sentiments, I briefly considered a liver transplant. But then I rejected it, based on my trust in the conventional wisdom, along the lines of "prolonging the life of an aging pet never justifies an expensive intervention." I saw nothing special in our situation to contradict that understanding, so I sadly had Scamper put down.

4.4 Decomposing a decision into stages

4.4.1 Going through "HOOPS"

A general-purpose way of looking at any decision dilemma is to ask yourself five generic questions: "What am I inclined to do? What can I do? What might happen? How would I like it? So what?"

Your answers correspond respectively to **H**olistic judgment, **O**ptions, **O**utcomes, **P**references and **S**ynthesis, abbreviated by their initials to HOOPS. I call this reasoning drill "going through HOOPS". I find this almost always useful, on major decisions, at least briefly, if only as an organizing framework for systematic reasoning.

H: Holistic judgment. What does your "heart" tell you to do, on the basis of unreflective intuition, or at least not with deliberate consideration. — i.e., a snap judgment. (♦ Joe feels a strong urge to marry Lulu, vaguely aware of her general popularity among his crowds.)

O: Options. Options are candidate immediate actions that are to be compared. Often, one comparison is an omission, i.e., not doing anything. Options can be *primary* actions (♦ Invite Lulu to the prom) or *informational* actions (♦ Check out Lulu's reading list first). An action can be a *one-step commitment.* (♦ Ask Lulu to elope on the spot), or a *provisional commitment* (♦ Ask out Lulu now; if it doesn't work out, then see if Jane is still available; otherwise, propose to Lulu; and so on). This is embarking on a *strategy*, only the first step of which is an irreversible action. What follows the action is merely an *intention*, for now.

O: Outcomes. The *outcome* of an action is all the *possible* events that follow it. Outcomes that are directly affected by an action are its *consequences*. An uncertain possibility (present or future) that is not itself affected by the action, but does affect your *satisfaction* with a consequence of the action, is a *setting* possibility. (♦ An economic slump would not be a *consequence* of marrying Lulu, but if its occurrence would make life with her worse than with Jane, it would be *setting*.)

P: Preferences are your value judgments. (♦ Joe values attractiveness above intelligence, but honesty above beauty.)

S: Synthesis. Synthesis is putting together your Holistic judgment with what you have judged about the other four HOOPS stages, and with any alternative judgment of the same dilemma, in order to make a choice (♦ Joe's (holistic) intuition favors Jane, but he puts more weight on his more structured deliberation, which favors Lulu. On balance, he favors Lulu.)

4.4.2 Caution — Confusing factual and value judgments

It is critical to keep the logical distinction between HOOPS components clear. Confusing HOOPS stages, especially outcome with preference, is a common source of unwise decisions. Misguided action often results from failure to distinguish facts from values.

♦ *Planning:* You are planning your evening. You are inclined to go to a movie. You think what else you might do: study for tomorrow's exam or catch up on your sleep. If you don't study there is a good chance you will fail some course; the other options won't have any longer term consequences. Failing the course could be a disaster; it could get you kicked out of school. Not worth the risk; so study. This risk trumps your initial inclination to go to a movie; so you study.

♦ *Terrorism case.* After the 2001 Twin Towers attacks, a government spokesman referred to the suicide bombers as "cowardly". Bill Maher, host of the TV talk show "Politically Incorrect" commented "Whatever else they are, they are not cowards". He was immediately taken off the air, on the grounds of his having expressed pro-terrorist sentiments, i.e., a *preference* judgment. In fact, he was making a *factual* judgment. He may have been mistaken in saying the bombers were not cowards, which may make him a poor judge of character, but his remark in no way implied approval.

♦ *Political correctness case.* Australian authorities canceled a Beatles tour in the 1960s, on the grounds that John Lennon's values were objectionable, after Lennon claimed that more people knew his name than Christ's. He was merely expressing a factual (if possibly mistaken) value-neutral judgment. The Pope might sadly agree with him, without outraging the faithful.

4.4.3 Implementing HOOPS

HOOPS can be implemented by addressing informally each stage in turn, often iteratively, and absorb the action implications naturally, without deliberate effort. That may be the only use of HOOPS you need. You can focus on a single stage, such as uncertain outcomes or preference, and handle other stages informally.

A decision tree, which graphically lays out considerations addressed in HOOPS (without numbers), can be used to informally structure your decision process. As elaborated in Chapter 5, you can quantify a decision tree by translating your uncertainty into "probabilities" and values into "utilities". (It involves estimating the utility of options, more specifically by comparing the utility of their possible outcomes, weighted by their probabilities.) This refinement of your analysis introduces a major new layer of analytic challenge.

4.5 Exercises

4.5.1 Informal reflection

Noah's Ark example. You are on a rescue mission to the tropical island — previously a mountain — Ararat, ravaged by floods. Dr. Noah and his ark of animals are sinking fast in the raging waters. Nearby, a leaking container with the last remaining pair of fabulous Uzlem birds is bobbing about. You must choose between saving either Noah or the birds.

a) Which do you try to save? State why informally.
Are there any critical questions you would need to address before making up your mind, including any not mentioned earlier? (Make any necessary plausible assumptions about Noah's case.)

b) You spot a jutting dry rock that Noah might just be able to get to in time to save himself. State informally what you would do now.

c) Before you can act, Noah cries out, "For nature's sake, save the birds! My life's work is done." What do you do now? How does part b) change now?

d) You save the birds, and Noah drowns. You discover both birds are male, so Uzlem birds are now extinct. Noah's family sues you for negligent manslaughter. You wail, "My God, I made a terrible mistake!" Did you?

4.5.2 Asking the right questions

1. In 2015 football star Chris Borland announced he was abandoning a lucrative athletic career on account of long-term health risks in the game. What questions would you have asked him to mull over, before he decided?

2. In the movie "Ransom", Mel Gibson's son is kidnapped and $2M ransom is demanded. Instead of complying, Gibson's character goes on TV and refuses the demand. He offers a bounty of $2M on the kidnapper, dead or alive, which he will withdraw when his son is returned unharmed. If you were the father, what issues would you address in deciding whether to pay the ransom or offer the bounty, if those were the only options you considered?

3. *Movie analog.* Think about a life-story you are familiar with that includes the protagonist's life up to your age. Would you trade with him or her during their period of life that corresponds to your age now? Why? How about their whole lives, compared with yours, including your own unknown future

4.5.3 Other exercises

Final project

What is the most likely aftermath if your proposal is adopted and if it is not? Describe each in two lines or less, compared with the current situation continuing indefinitely — which may or may not be the same as for the "no" option. Which of these two outcomes would you prefer if it were certain? How, if at all, is your choice affected, given any uncertainty about either option outcome? Does this choice differ from your initial intuitive choice?

Energy policy

Suppose there is a bill before Congress to abolish nuclear power: specifically, no new reactors are to start operations and all the 100+ commercial reactors in the US are to stop operation by Year 2050. Off the top of your head, would you support the bill, without any further consideration of what it would entail? Be prepared to discuss and evaluate more deliberately — but qualitatively — the pros and cons.

Racism

Imagine this argument: "At the risk of upsetting Muslim friends, I must speak up for what has been called 'profiling' in airport security screening — i.e., focusing attention on certain ethnic and religious groups. To be sure, it selectively inconveniences innocents, which is certainly to be regretted. (Why not compensate them for it? $10 per 'friskee' would surely help calm indignation.) But profiling also promises to make investigative effort more efficient in uncovering the guilty, and that is surely the dominant concern."

"I am not necessarily a racist to want to use any available clues to target an effort where the guilty are most likely to be found. The clues surely include ethnicity and religion (along with age, physique, behavior, etc.), particularly if I observe that most terrorist attackers have a certain profile, at least in cases of international terrorism. If all I know about a passenger is that he is Muslim, that will raise the chance (however small) I assess that he is a terrorist. We should surely look more carefully at a fit young Arab with a one-way ticket than at an elderly nun en route to Lourdes. Picking suspects in the theft of a menorah, I would search Jews before gentiles. Likewise, when predicting a mass shooter or domestic terrorism, I would profile for white men."

On what grounds might someone reasonably disagree with this position?

Veggie burgers or Key Lime Pie?

You are feeling peckish in mid-afternoon and find veggie burgers and key lime pie in the fridge. Discuss informally which you should eat, in HOOPS terms, max one line per element. Here is a sample response.

H: (holistic). I would like veggie burgers because it's healthier and I generally like to be healthy.

O: (options). 1. eat veggie burger, 2. eat Key lime, 3. eat both, 4. eat neither.

O: (outcome). If eat the burger, I will have a more full, less fat feeling. If I eat both, I'll be too full. If I eat neither, I'll still be hungry. If I eat half of both, it may balance hunger and craving.

P: (preferences). I prefer to be healthy and put off gratification in terms of feeling fit (rather than have immediate gratification of sugar high).

S: (synthesis). I will combine all of the above intuitively.

Whether or not to swim to shore from a cruise ship

Your cruise ship has hit an iceberg and you are inclined to believe the captain's assurance that there is nothing to worry about. The shore is close enough that you are quite sure you could swim to it safely, but it would be a most unpleasant experience. Do you think you would try? What HOOPS component(s) would it mainly depend on? Here is a sample response:

Decision depends on:

1. Are there safety boats?

2. Am I relatively sure that I could choose to swim later if it looked necessary?

3. Do I trust the captain? (Have I had a good sense of him so far and his competence?)

4. Do I know of cruise ships that have sunk in similar circumstances?

5. Do I swim well?

6. Is there anyone on the boat who I need to take care of?

7. Is the water cold and could I likely survive the swim?

8. Am I safe on the boat right now? (Do I see water leaking in?)

9. Do I value current comfort over possible future discomfort?

10. Will I sway toward not swimming because I simply don't want to be cold, even if there is a greater chance I will die if I stay on the boat?

4.5. EXERCISES

Decision: I would stay because I prefer the comfort and knowledge of the status quo.

Political disagreement

Pair with someone, e.g., another student, with whom you disagree on a topical public policy option — say: whether the US should intervene in "Islamistan's" civil war. State briefly what you think the sources of your difference are:

Options   what else, if anything, the US would do;
Outcomes   what would happen in either case;
Preferences   which consequences of an option you favor;
Synthesis   how you reason from these considerations, or something else (e.g., you follow a different "party line").

4.5.4 Distinguishing factual from value judgments

Which of these statements imply judgments of: choice (or advocacy); a factual assessment of possibility (perhaps uncertain or wrong); a value judgment of preferences; or are not clear? Indicate "C" for choice; "P" for possibility; "V" for value judgment; or "U" for unclear. (Add a brief explanation only if you think it necessary.)

1. "Al Gore would have become the President of the US if all the ballots in the State of Florida had been recounted."

2. "At least 40% of our company's workforce must be female."

3. "Animals in experiments have rights, but saving human lives is more important."

4. "The rich should have a lower tax rate than the poor, as reward for their superior contribution to society."

5. "A 'foot and mouth' outbreak cannot be contained by vaccination."

6. "Sustained economic growth should always take precedence over the pollution reduction."

7. "Letting middle-eastern immigrants into England without limit will produce rivers of blood (violent conflict)."[1]

8. "A woman's place is in the home."

9. "In negotiations, it helps your case to misrepresent your values."

10. "Prospects of extreme punishment encourage child abusers to murder their victims."

11. "Child abusers should be castrated."

12. "Animals in experiments have rights, but saving human lives is more important."

---

[1] Enoch Powell, British politician 1977, whose prospects of becoming prime minister were destroyed by this statement.

Chapter 5

Hip surgery case

The following personal case illustrates primitive PDA, including ADT applied to a real dilemma.

In 1989, at age 55, I had to decide whether to immediately have an operation to replace a painful arthritic hip — "CUT" — or to delay surgery for now — "WAIT". Table 1 shows my initial thoughts.

Table 1: Impressionistic evaluation of choice: What to do about bad hip.

Decider's goal for this choice: To be more active without too many immediate problems.	
Option: CUT (Regular "sawing" operation now)	*Option:* WAIT (Wait until hip gets much worse)
Possible Outcomes: Successful operation almost certain	*Possible Outcomes:* Increasingly miserable next ten years
I'll be in good shape after an unpleasant experience	A better treatment may turn up
Cost is affordable	Something will need to be done eventually (if I do not die first)
Other options: Seek to have "resurfacing" operation done now Preferred option: CUT	

5.1 Preliminary evaluations of hip choice

Before undertaking any significant analysis, I made a few preliminary evaluations of my choice:

Why I think CUT is the best decision: It will improve the quality of my life significantly for the next five years or so, while I am still active. I can stand the cost, pain and inconvenience of surgery right now and, anyway, I will have to have it sooner or later. This operation seems quite safe, and although the new resurfacing operation might bother me less, the doctor is not sure about it and does not recommend it.

5.1.1 Intuition

My first intuition was to WAIT. I had a visceral revulsion at the thought of my thigh bone being sawn through, which over-powered the appeal of reducing later discomfort.

5.1.2 Consultation

My surgeon thought I was a bit young (55) for this type of operation and firmly advised me to WAIT a couple of years which apparently was consistent with general medical "guidance".

I was not concerned about him having any professional bias, since he was advising counter to his financial interest. I was inclined to trust his WAIT advice.

5.1.3 Cognitive vigilance

On reflection, I recognized that, like many others, I tend to over-value immediate over delayed satisfaction. This tipped the balance toward CUT.

5.1.4 Analogy

Then I considered what I could learn from the comparable hip surgery experiences of other people that I knew. I had very little to go on, but my impression was that those who had it done earlier were glad they had not waited — or at least not did not regret it. This was weak evidence, but it encouraged me somewhat to go ahead and CUT.

5.1.5 Visualized aftermaths

I visualized the most plausible aftermath scenario with and without an early hip replacement, almost as a movie of my life following each option, bearing in mind the possibility of aftermaths both better and worse than the most plausible. *I deliberately gave it more weight than my reluctance to CUT. I thought: "Surgery will be more inconvenient and unpleasant now, but that is really less important to me than later satisfaction, so I prefer to CUT."*

I imagined that WAIT would lead to progressively more pain and difficulty walking for another ten years; then I would operate which, with improved hip replacement technology, would reduce the short-term negatives of surgery and avoid a second hip replacement. CUT would involve a very disagreeable short-term surgery experience, followed by mainly pain-free and mobile movement; until it deteriorated to the point where some 20 years later, I would need another hip replacement. Its impact would be similar to an immediate CUT, but probably terminated by my death some years later.

The actual aftermaths could be different from these "best guesses", but that uncertainty didn't much affect my comparison of options since it was comparable for both options. I preferred the life that I imagined following CUT with moderate confidence.

5.1.6 Interim synthesis of evaluations

These four preliminary evaluations thus had conflicting pulls on whether to CUT or WAIT, but intuitively CUT seemed to win out.

5.2 Main evaluation effort

5.2.1 Student participation in hip choice

Class term project

It so happened that I was starting to teach a short course in decision skills to a class of local 8th graders. I arranged with the surgeon to delay the surgery decision for the two-month duration of the course. During that time, I used this hip surgery choice as a case study on which students exercised what they were learning. The students acted as decision consultants to me, as their decider-client.

As issues arose, they had my surgeon answer questions of fact (i.e., his medical judgment) through me, and they asked me for my value judgments (i.e., what was important to me). The class project consisted of essentially picking my brains — and indirectly the surgeon's — according to the decision skills they were learning.

The weekly class assignments extracted my judgment sequentially, class by class: first, specifying my options; then assessing option outcomes qualitatively; then eliciting quantitatively my preference for options (represented by pluses/minuses for advantage/disadvantage for WAIT option on each goal, as shown in Table 2). The preference judgments came directly from me; the outcome judgments were based on my conferring with the surgeon.

I led the class through a hybrid decision strategy over the two-month period of the course. We looked at my choice several different ways and considered them together to draw a conclusion. At the end of the two-month course, we took a vote and nearly all students recommended CUT.

A decision scientist colleague served as teacher to facilitate class discussion. She had them ask me questions prompted by their homework assignments.

My adaptation of class project

Based on what I learned from class discussions (which was by no means negligible), I adapted their analysis to reflect my own best judgment. I registered my appraisal in the same format as they had, but modified their specific input, specifically their outcome criteria and my preference comparison of the two options.

5.2.2 Simplifying the options

To make the problem more tractable to analyze, I first treated my choice as if my only WAIT option were to delay ten years. In fact, I could reconsider surgery at any time from now on (say, in the event of advances in hip surgery technology). This simplification penalized the WAIT option, because I could

5.2. MAIN EVALUATION EFFORT

actually change my mind, whenever changing circumstances indicated. I allowed for the model's inflexibility later, before actually making my choice.

I also tallied pluses and minuses, representing how much better or worse CUT was than WAIT.

Table 2: Pro–con tally of options for hip choice: CUT–WAIT difference.

Criterion	Consequence comparison	Difference
Near-term satisfaction	Ten years improved comfort, mobility, etc.	+ + + + + +
Longer-term satisfaction	Negligible difference	0
Near-term cost	$2000 for operation ten years early	− −
Immediate satisfaction	Horror of bone sawing, post-op pain	− −
Delayed pain	Same PAIN ten years later	− −
Immediate death	Negligible chance during surgery	0
Risk if re-operation	Moderate vs. negligible chance	− −
Net difference		+
Adjustment for model mismatch		0
Adjusted difference		+
Preferred option: WAIT		

5.2.3 Choice criteria and option impacts

As shown in Table 2, I identified the six criteria shown and assessed how the impact on each of CUT differed from WAIT. I evaluated the importance of the difference as a number of pluses or minuses, noting whether they favored the CUT main option or not with a plus or minus.

The most important difference, by far, was in the "near-term satisfaction" of the next ten years' comfort. I predicted that CUT would improve the next ten years for me by dramatically reducing pain and permitting me to engage in a wide range of activities, which WAIT would deny me. Accordingly, I gave it six pluses.

My preference was checked by comparing it with the minuses for the other offsetting criteria.

Adjustment for model mismatch

The model differed from my perception of reality in two significant ways. I treated the WAIT delay as exactly ten years, whereas I could CUT at any time earlier or later. This increased flexibility favored the CUT. On the other hand, I was restricting the impacts to an integral number of pluses and minuses and rounding up. This exaggerated the impact of small effects, which, in this case, favored WAIT. These two mismatches favored different options, and I considered that they canceled out so I made no net adjustment for model mismatch.

I was originally inclined to give CUT only three pluses for "short-term satisfaction", but then recognized a general tendency of people to underestimate the larger differences between options. So, I increased the pluses from three to six. As an additional check on my scoring, I took note of the fact that I scored the "near-term cost" of $2000 cost of surgery to one minus, implying that I valued short-term satisfaction at around six times $2000 or $12000, which felt about right.

Totaling pluses and minuses gave CUT a slight edge at one plus.

5.3 Synthesis of conflicting appraisals

5.3.1 Synthesis at the time

In all, I had evaluated my choice five different ways, with the following results (noting in parenthesis, how much confidence I had in each evaluation):

- Consultation: WAIT (high)

- Intuition: CUT (low)

- Analogy: CUT (medium)

- Plus-minus tally: CUT (high)

- Plausible aftermaths: CUT (medium)

All but one of these five appraisals pointed to CUT, with either low or medium confidence, which would normally leave me with high confidence that this CUT consensus was wise. However, the exception was the surgeon's advice that conventional medical wisdom pointed to WAIT with high confidence. How to reconcile this marked disparity?

I presented the CUT choice and the reasoning behind it to my surgeon, who had advised me to WAIT. On reflection, he acknowledged that he had not originally taken everything we had into account, changed his mind, operated, and I had no more pain. The surgeon accepted my contrary input judgments as reasonable.

Combining the five appraisal methods choice left me with a comfortable feeling that CUT was the right thing to do. The surgeon agreed and performed the CUT operation with every indication of success. From the perspective of 20 years later, I have no reason to think I made a poor choice.

5.3.2 Hindsight synthesis

It was not until some years later, when I revisited this case, that it dawned on me that the surgeon's advice to WAIT was predicated on value judgments shared by him and the medical community, but different from mine. A significant element in my choice was the fact that, since I had health insurance, the operation would not cost me anything extra. The surgeon, as a responsible citizen, would presumably take that cost into account because that would be enough to switch the balance. So, making allowance for that consideration, the consultation input did not, after all, conflict with the four approaches that favored CUT.

The surgeon did not bring this issue up when we discussed the disparity in our choice prescriptions. I suspect he was not really persuaded by my argument, but bowed before (or was intimidated by) the apparent authority of my supposed logical expertise.

5.4 Seek more information before deciding?

Before I decided to CUT, I could have read up on relevant medical literature. However, I reasoned: "With the little I know now, I may regret my CUT choice. However, I cannot learn much that would change my mind and what I could learn would not be worth the trouble."

I was pretty sure that such research would not change my choice, so I didn't bother to seek more information.

5.5 Post scripts

5.5.1 Sequel

End of story? Not quite.

The surgeon advised me to go very gently on the hip for several months and I agreed unhesitatingly. All was well for a couple of weeks, but I was so exhilarated at being pain-free at last, that I took a high-impact aerobics class. Of course, my hip gave way; the metal implant was dislodged and I limped for decades after.

5.5.2 Action vs. decision

The moral: "Thinking smart is not the same as acting smart." I had made the right decision "to go gently". But, through human frailty, I didn't act on it and paid for it dearly.

The long-term outcome was worse than it could have been, because I acted foolishly by disregarding my best judgment. However, I have no reason to think that WAIT was the wiser choice. Being aware by hindsight — or even in foresight — that I am prone to suffer from dumb subsequent actions did not invalidate the CUT vs. WAIT decision process.

Note that the chance I assess for re-operation if I CUT depends on whether I avoid jolts. I was confident (misguidedly) that I would avoid jolts, which made my chance of re-operation very low, implying that overall, re-operation was quite unlikely. Quantification: Conditional on avoiding jolts, probability of re-op, 10%; re-op conditional on jolts, 50%.

5.5.3 A later choice

Fast forward 24 years. On the eve of my 80th birthday, I was faced with a choice similar to the one I faced at 55. The replaced right hip became

5.6. EXERCISE

again painful and debilitating to the point that I needed a wheelchair to get around. I could again replace it, with the prospect of an operation more troublesome than before, because of my advanced age, but promising largely restored comfort and mobility thereafter. (Note that the prospect of a re-operation like this figured in my original dilemma as a negative element in my long-term satisfaction criterion.)

By analogy with my earlier hip replacement dilemma, a con for CUT is the operation itself will be a worse experience, but a pro is that it should improve the quality my life more thereafter. However, another con is that "thereafter" will be much shorter. At age 80, I believed my "life expectancy" was around six years with WAIT, and perhaps a year less with CUT. I judged that a poor "return on investment" short-term pain and decided to WAIT. (Analogy: I wouldn't get new tires for an old car that I expect to scrap shortly!)

As it happened, I shortly suffered an accident, my hip took a major turn for the worse, and I had no choice but to CUT the day before my 80th birthday.

5.6 Exercise

- In this case, state briefly what you think my answers to the four HOOPS questions are, for the original decision.

- If I were to learn that a new hip surgery technique, known as resurfacing, was about to be approved, which had faster and less painful recovery, how do you think that should affect my original CUT/WAIT choice?

- If I had thought harder about my choice and assessed an appreciable probability that I would imprudently seriously jolt my hip (as in fact I did), should that consideration affect my CUT/WAIT choice?

- Suppose I knew that, in five years, I would be able to take a wonder drug that would painlessly, permanently restore my leg to full performance, but it would cost twice as much. Take a guess at what that would change in my pro/con tally chart for WAIT vs. CUT.

Chapter 6

Quantitative ADT modeling

This chapter deals with the basics of converting qualitative decision reasoning into quantitative models according to the logically consistent rules or norms of *applied decision theory* (ADT). These norms make use of quantitative readings of parts of your mind that judge uncertainty (probability) and value (utility). They allow checks of the consistency of these readings, and they can identify the choice these readings imply. (How to use ADT models effectively on real dilemmas is addressed in Chapter 8.)

6.1 Quantitative decision reasoning

6.1.1 Reasoned decisions

A decision process that goes systematically beyond *intuition or holistic* judgment can be said to be *reasoned*. Qualitative versions of reasoned argument were discussed in Chapter 4. Quantitative versions usually consist of making the same arguments more precise and more defensible, in the form of a numerical *model*.

6.1.2 Models

A *model* is an entity substituted for another that it resembles in some major respect, but not in irrelevant others (just as a model airplane resembles a real airplane in only some respects). An analytic model resembles a decider's reasoning, but not her physical properties. The argument is logically coherent. In effect, it converts decision "art" into decision "science".

6.2 Applied decision theory (ADT)

6.2.1 Normative decision theory

Normative statistical decision theory comprises a body of logical norms or rules that check whether some of your judgments of uncertainty and value, expressed as numerical probabilities and utilities, are coherent. Illustrative decision theory norms: Probabilities of exhaustive and exclusive events sum to 100%. The utility of an event situation or option equals the probability-weighted average of component utilities.

6.2.2 Applied decision theory models

Applied decision theory (ADT) is a class of *prescriptive decision analysis (PDA)* tools that obey decision theory norms. Any *sound* reasoned argument can be quantified by ADT.

This involves you taking numerical readings on your mind-contents — for example, as "probabilities" and "utilities". It involves creating quantitative *models* of your judgments about a choice, which you treat as substitutes for relevant parts of your mind-contents, from which you infer your preferred choice. A model quantifies your judgments of options, outcomes and preferences[1] and draws inferences from them according to decision theory norms.

My philosophical position is *personalist*, in the sense that analysis refers to a specific person's judgment. This perspective is appropriate to a decider (in this case, yourself) attempting to use her judgment to resolve a particular dilemma. (By contrast, an *impersonalist* perspective avoids using input specific to any particular person.)

You can evaluate ADT (or other) decision-aiding approaches by estimating how close they come to some *ideal* unknown judgment, which takes account of all that you know and feel.

---

[1] Or, at least, judgments that you accept, if not originate.

6.2. APPLIED DECISION THEORY (ADT)

6.2.3 Utility as a measure of satisfaction

Your satisfaction with an option can be measured by a *utility* number between 0 and 100, whose ends correspond to some arbitrarily specified bad and good state, respectively.

♦ As a marriage partner, bachelor Joe puts Lulu's average utility at 71.0 and Jane's at 65.5 on a scale from 0 to 100, corresponding to his evaluation of certain bad and good marriages he is familiar with. Implication: to be consistent with these evaluations, he should prefer Lulu.

Getting the *absolute* scores right is less important than getting the *difference* between options right. If all scores are five points too high, that will have no effect on which one is highest.

♦ Joe is less sure that 71 and 65.5 are right, than he is of the difference of 5.5 between them, which is what matters in choosing between them.

The logical relation among all your utilities and probabilities derives from *decision theory norms*, a logic for making sound decisions, developed about 100 years ago. It involves breaking a decider's reasoning into separable judgments, and putting numbers to those pieces, and calculating the choice they logically imply, according to a model obeying certain logical norms (such as "probabilities of exclusive and exhaustive events must add to one").

6.2.4 Uncertain utility

Locating options on a utility scale will typically be subject to uncertainty. This can be measured by a *probability* number, between 0% and 100%, corresponding to impossible and certain, respectively.

Your utility for an option may depend on uncertain possible developments. The overall utility of an option is the expectation of its utility. This would be the average utility if the probabilities represented the relative frequency of outcomes, in the unrealistic scenario in which the decision happened hundreds of times.

You should logically value most the option whose outcomes you judge to have the highest *average personal utility (APU)*. The average is a weighted average. The utility of each outcome is multiplied by its probability, and the results are simply added to get the average for each option.

Within the ADT format, present a compact kit of decision tools, to be used singly or in combination. These tools tap into different parts of your *mind-content*. They are suited to a wide variety of levels of decider intellectual maturity and training and involve no math beyond simple arithmetic. But they need a good deal of thoughtful effort to be really effective. ADT modeling of a choice so that it improves on unaided judgment requires skill and training. Although ADT is grounded in advanced mathematical statistics, the tools can be in the form of simple numerical models, using no more than elementary arithmetic. Moreover, they will not replace your regular decision processes — only complement them — and then not always. However, although the models themselves are simple, the thinking that goes into them is not always so simple.

The model will often (but not necessarily) correspond to a HOOPS formulation (see Chapter 4). The essence of any ADT is to ask the right questions and to know what to do with the answers. The latter may be simply to let the answers sink in and have your intuition or regular reasoning processes take over.

6.3 Quantifying HOOPS: Decision trees

A major class of ADT reasoning is *projection*, whereby the possible outcomes of options are modeled and *utility* is attached, depending on some uncertain condition.

The options considered are branches of an initial *act fork for alternative options*; followed by one (or more) *event forks*, labeled with possible outcomes with their probabilities. The utilities that result from the combinations of preceding options and possibilities are on the right. The *subjective expected utility* (SEU) of possible outcomes is calculated as the average of their utilities, and this is the resulting utility of the options.

6.3.1 Basic decision tree mechanics: Simple example

The decision tree is the graphic version of a model of the APU of options, commonly used for choices involving major uncertainty. The preferred option is derived from processing analytically certain facts and preferences.

The formal procedures of decision tree analysis can be demonstrated in the context of simplified examples. The simplest case is a gamble where money is D's only criterion, can be taken as a surrogate for utility, you are not averse to risks and uncertainty and assessment is uncontroversial.

♦ I offer you a gamble, where I flip a coin. If it comes up heads I pay you $200, tails, you pay me $100, do you accept? It is easy to see that your average probable return from this gamble is a gain of $50, the mid-point between $200 and −$100. So, if you are prepared to "play the averages", you accept. That is the average amount you would win each time if the game were played thousands of times (not that this is a good idea). It is also the result you get from multiplying each amount by its probability (.5) and adding the two products.

6.3.2 "Bachelor Joe" example

Figure 2 shows a decision tree, in the context of a more interesting, but still very hypothetical example. Bachelor Joe is deciding whether to propose to fun-loving but high-spending Lulu or career-focused, sensible Jane. The tree is a model of Joe's perception of his marriage problem. It follows common decision tree conventions.

In this very simplified example, shown in Figure 2, Joe's evaluation of Lulu depends on whether or not she gets a high-paying job, which he assesses at 80%. If Lulu gets a good job (probability 80%), Joe scores her utility at 75; if not, 55. This implies a SEU marriage to Lulu of $(.80 \times 75) + (.20 \times 55) = 71$. Given that his utility for Jane, calculated as $(.10 \times 25) + (.90 \times 70) = 65.5$, this implies that his wise choice *based on these inputs*, would be to marry Lulu. Notice that the better outcome with Jane has a higher probability than the better outcome with Lulu, but that is not enough to make up for the difference in utilities.

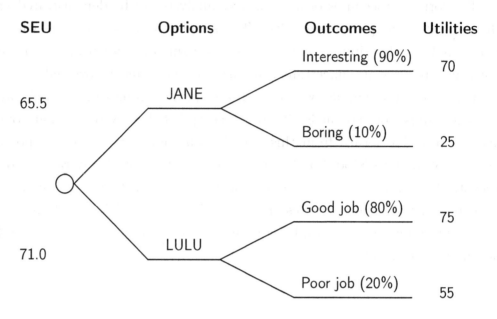

Figure 2: Decision tree for bachelor.

The utilities scale is from 0 to 100, where 0 is Joe's evaluation of a neighbor's dreadful marriage and 100 is his evaluation of the blissful married life of a couple he knows. Jane's utility is the probability weighted average of the two possible outcomes of marrying Jane, 65.5. The same logic applied to Lulu produces a utility of 71. This is higher than Jane's 65.5, so, on these judgments, Joe should choose Lulu. Notice that the probabilities matter. The un-weighted averages are 65 for Lulu (the average of 55 and 75) and 47.5 for Jane, which would favor Lulu by a larger amount.

The options considered are branches of an Initial *act fork for alternative options*; followed by one (or more) *event forks*, labeled with possible outcomes, each with its probability. Two immediate options, JANE and LULU, are represented as branches of a *choice fork* with a circle at the base, and branches are the options, labeled in capitals. (Joe could consider many more options than the two shown, as additional act branches.) The outcome of each option is shown as branches of an *event fork*. Branch probabilities (shown in parentheses) depend on the preceding option, shown to the left. Thus, if Joe opts for JANE, his probability of finding her boring is 10%.

6.4. ASSIGNMENTS

The utilities that result from the combinations of preceding options and possibilities are on the right. The *subjective expected utility* (SEU) of possible outcomes is calculated as the average of their utilities, and this is the resulting utility of the options. Each outcome utility is multiplied by its probability, and the results are added up for each of the options. The utility of any sequence of option and possibility is shown in a column to the right of the corresponding two-segment tree *sequence*.

6.4 Assignments

1. Consider the following. Suppose I offer you the following gamble: You pay me $200 to play. The result depends on how the maximum temperature next Monday compares with the maximum temperature today. If the temperature is higher, I pay you $2000; if it is lower, you fail a course (or go bankrupt, if you are not taking any courses). If they are same (measured to the nearest degree Celsius) you get nothing. Would you accept? Draw and analyze a decision tree that confirms your choice. Which of the original assumption(s), if wrong, might change my decision? Why?

2. Tomorrow, Dostoyevsky will pay me the $1200 he owes, when he cashes his uncle's check, but only if he survives that Russian roulette that he is first going to play.[2] Before he plays, he offers me the $900 he has in his pocket to cancel the debt. (Unless I accept this offer, I will not get anything before he plays, and nothing afterwards if he "loses".)

Making any necessary additional assumptions, but accepting as true everything stated above, calculate whether I should accept the $900 now, or wait till after Dostoyevsky plays and hope to collect all $1200 (which I will only do if he survives).

3. Consider the following:

a) I offer you a deal. I will toss a coin. Heads, I give you $2; tails, you give me $1. Do you take it? Draw and analyze a decision tree that confirms your choice.

b) Now I offer you a new deal. Heads, I give you $10,002; tails, you give me $10,000. Do you take it? Draw and analyze a tree for this deal.

---

[2] In Russian roulette, the player shoots himself with a revolver which has six chambers, only one of which contains a bullet.

Chapter 7

Family case study: C-section vs. "natural" child-birth?

This dialog refers to a personal three-generational family episode of mine in 1998. Note to reader: I use the term "natural" to mean a vaginal birth, with or without pain relief.

7.1 A baby delivery dilemma: Natural child-birth or C-section?

Karen: Dad..., I know you usually work with managers and such, but could you help me make a very personal decision?

Rex: Well, I'll try. The principles are much the same.

Karen: You know I am about to give birth to boy-girl twins. Well, the doctor says there's a dilemma. Lucy, who will be first, is positioned fine for a natural delivery; but Sam is breach — which means feet down. Unless he can be "turned" after Lucy, I will probably have to have a C-section for him. Should I wait and see what happens with Sam at delivery, or go ahead and have the C-section for both births preemptively?

Rex: How do you feel about it now?

K: I'm inclined to deliver Lucy naturally and just hope that I won't need a C for Sam. But I feel uneasy about it. If it turns out that I still need a C, I will be sorry I didn't do it from the start.

R: How sorry?

K: Either way, having a C is pretty difficult. But having a natural birth first is somewhat worse, because then I will end up going through both labor *and* surgery, which would be arduous and slow my recovery. Cost is not an issue, as I am covered by insurance.

R: Is there a good chance that Sam can in fact be turned so that both births could be natural?

K: According to the doctor, it could go either way.

R: Let me sketch out the gist of what you have just said. The two rows in this table correspond to your two options: C-FIRST and TRY NATURAL. The columns lay out what might happen in each case, how likely, and how you would feel if it did happen. Does organizing them this way help?

Table 3: Karen's informal evaluation.

What can I do?	What might happen?	How likely?	How would I feel?
C-SECTION FIRST	Double C-section	Certain	Pretty bad (long recovery)
TRY NATURAL	Two naturals	Even chance	Best case
or...	Natural then caesarean	Even chance	Worst case (long recovery)

K: I suppose so. Just looking at the table does seem to confirm my intuition of having a natural delivery first. But I'm still not quite convinced, and I'd like to be able to review the argument constructively with my husband, Sean.

7.1. A BABY DELIVERY DILEMMA

R: Let's try to put numbers on the pieces of your reasoning in a simple decision theory model.

R: First, your uncertainty about turning. Can you get your doctor to put a probability on being able to turn Sam?

Later, after consulting with doctor:

K: She still won't be committed. She supposed it was about as likely as not, but she couldn't go further than that.

R: OK. Let's call it 50% for now.

Now, about your preference for outcomes. On a scale from 0 to 100, where 0 is the worst (natural-then-C) and 100 the best (two naturals), how happy would you be with the one in-between (double-C)?

K: That's a tricky one. A double-C is certainly nearer to the worst outcome, natural-then-C, than it is to the best outcome, two naturals. So a double-C would be nearer to zero than 100. Let's give it a 20.

R: We can now run the numbers, to see what these judgments of uncertainty and preference mean for your choice.

The C-First option guarantees you will have a double-C outcome, to which you just gave a 20. If you TRY NATURAL, there are two possible outcomes, natural-then-C at 0, and two naturals at 100. So you know the average is going to be somewhere between 0 and 100. We get an exact number by multiplying the probability of each outcome (.5, .5) by its preference (0, 100). Your average preference for TRY NATURAL (as shown in Figure 3) is therefore $(0.5 \times 0) + (0.5 \times 100) = 50$. This is higher than a certain 20 for C-FIRST and so preferred.

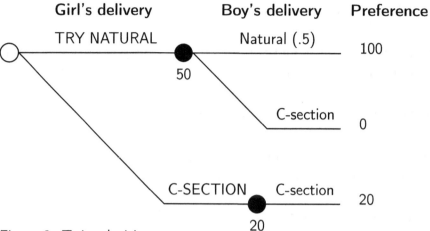

Figure 3: Twins decision tree.

Figure 3 shows this argument in the form of a very simple "decision tree". The numbers below the solid circles evaluate the options: 50 for TRY NATURAL and 20 for C-FIRST. They are the result of multiplying and adding probabilities in the middle column, given the option of the possibilities leading to the preferences shown on the right. Only TRY NATURAL has uncertain outcomes quantified.

Since 50 is better than the 20 for C-FIRST, you should apparently TRY NATURAL, and by quite a bit.

K: Why do you say "apparently"?

R: Well, "Garbage in, garbage out" they say. Perhaps the numbers you gave me are not very solid. If they change, so might your conclusion.

Realistically, could your probability of turning Sam drop below the 50% we used? If it were less than 20%, your average preference for TRY NATURAL would be lower than 20 ($.2 \times 100$). So it would then be even worse than we first thought, and therefore certainly out.

7.1. A BABY DELIVERY DILEMMA

K: I don't think my probability could drop that low. In fact, I think the doctor is *more* confident than 50% that she can turn Sam. She might have said "it could go either way" just to be on the safe side.

R: So the argument for TRY NATURAL would be even stronger.

How about preference? Could your preference for ending up with a double-C be closer to two naturals (the best) than to natural-then-C (the worst)? That is, instead of the 20 you gave originally, could your preference for a double-C be actually above 50 (i.e., nearer to best than to worst)?

K: Definitely not!

R: Then plausible shifts in your basic judgments wouldn't change your choice.

But before you decide, let's look at something else. Perhaps you don't like to take risks. With the C-FIRST option, you are quite sure of having a double-C, which is quite bad. With the TRY NATURAL option you have a 50:50 chance of *either* natural-then-C deliveries which is even worse, *or* two natural deliveries, which, of course, is best. Do you want to chance it?

K: I think so. Even though I don't like risk much, the gamble looks so much better than the sure thing that I'd take it.

R: So, putting numbers on your judgments like this fits with your original intuition: that you should have a natural delivery first. That choice sounds solid, and not likely to change if you dwell on it more, with or without the numbers.

K: I agree and I feel a whole lot easier in my mind now about what to do. I'll go over what we've discussed with my husband. If he doesn't see anything wrong with it, I'll go ahead and try for a natural delivery first.

She did. Both babies were delivered naturally (and did fine).

7.2 Postscript (six months later)

R: So, you opted to try for a natural delivery-first, you were able to have both deliveries naturally and now have two bonny, healthy babies. Did you find our decision-aiding exercise useful?

K: It certainly had a good outcome, two successful natural deliveries. Without it, I was probably going to take the more conservative recommendation and have the C-section first. As it was, I felt reassured going with my intuition.

R: Bear in mind that you could have had a nasty surprise and had to have a C-section anyway. You could have made a misguided decision, but got lucky. Your happy outcome was a good sign of a wise choice, but not conclusive.

K: However, something is bothering me. I recently talked to an experienced obstetric nurse, who said that, in this particular case, it might have been wiser to do C-FIRST, even though it proved unnecessary in the event. Since Sam had not been gaining weight as the due date approached, she would actually have recommended an *early* double-C to safeguard his health. I had not even considered that option. If her judgment was sound, does that mean my TRY NATURAL choice was ill-considered, even though it turned out well?

R: Not if you made the best choice, given what you knew *at the time*, regardless of the outcome. If you'd had that new information then, an early C-section might have been the best choice. Perhaps you should have sought more expert advice (like the nurse's) *instead of* our doing the numerical analysis, if there wasn't time to do both.

K: When I am making similar choices in future, can you give me any general advice on how I should make up my mind?

R: When you have serious concern about making a bad decision, by all means consider complementing your normal thinking process with some modeling, with the help of a specialist (as you did here, with me). It should improve your prospects of a good outcome, if you don't over-weight the modeling in your choice.

7.3. TECHNICAL COMMENTARY

However, to be confident of at least improving on your normal unaided reasoning, I think you need to learn enough about the art of decision aiding to be able to take the lead in the analysis, and to use a specialist only as a technical resource. You need to be the one to determine what, if any, formal decision process to use (beware of experts pushing their specialty!); to verify that your factual and value judgments are accurately quantified; and to integrate model results into your informal thinking. To be able to do all this effectively, you would need decider-oriented coursework on applied decision theory (and/or apprentice to a practicing decision aider, if the opportunity arises). In any case, the training should educate your intuition.

7.2.1 Post-postscript

In 2019, both children are in college and doing well.

7.3 Technical commentary

7.3.1 Decision-making issues raised

This was a real and personal case, where I used a numerical model to help a family member make an important choice. It illustrates a number of decision-aiding issues, which we will take up more systematically later. They include:

- informal reasoning;
- quantifying the reasoning in a model, providing input to the decider;
- deriving output;
- interpreting the output;
- resolving uncertainty;
- accounting for preference and risk aversion;
- hybrid reasoning;

- taking action;

- useful exercise;

- good choices vs. good outcomes;

- knowledge vs. reasoning.

7.3.2 Applied decision theory

The primary approach to addressing these issues was ADT (applied decision theory). This involves quantifying judgments in the form of a numerical model. This is essentially an equation, where a *target judgment* is equated to a formula where the elements are relevant factual and/or value judgments, expressed as probabilities and utilities respectively. These are linked in such a way that they obey logical decision theory norms.

7.3.3 My contribution as analyst

My daughter was evidently content to rely on my guidance in this serious dilemma. However, she resolutely refused to allow these same twins to be in a car I am driving! She apparently accepts that my years of ADT training have benefited the deliberate decisions I make when I have the time, but, like all who know me well, she profoundly distrusts my response to sudden contingencies, such as dangerous traffic incidents.

7.4 Exercises

1. One obstetrician strongly urged Karen to have a C-section first, without addressing any specifics of her analysis. Given his superior medical expertise, is there any reason why she should not simply let his judgment supersede hers?

2. Rerun the decision tree numbers with Karen assessing 20% probability that the boy could be turned and evaluating the outcome of a double C-section as mid-way in desirability between "both natural" and "natural-then-C-section". How should it affect her choice?

7.4. EXERCISES

3. In the real-life personal problem, Karen had to decide whether to have both her twin babies by caesarean or to have the first, Lucy, naturally and take a gamble that she also could have the second, Sam, who might have been more robust, naturally. After this analysis, she opted for the latter option, managed to have both deliveries naturally, and both mother and babies did fine. Here are some answers:

- Did that good outcome vindicate the soundness of her choice?

    No, it could have been an unwise choice, in the light of what she knew at the time, but she got lucky.

- Later, the mother was given new medical research that a fetus in the "feet down" position, which Sam was, should almost always be delivered by caesarean, because of possible injury due to "turning" him. If she had believed this at the time, should she have taken the "double caesarean" option? How do you think this consideration would change the drawing of Figure 3?

    Yes. Based on this new knowledge, she would expect that double caesarean is best, whether or not Sam can be turned. The new information would change outcome utilities shown on the Preference scale. In particular, if the utility of two naturals (the top line) dropped below 40, then the expected utility of the choice to try natural would drop below 20, the utility of the double caesarean.

- Does finding out this new information indicate she made a mistake at the time?

    No. She did the best with what she knew at the time.

Chapter 8

Using ADT models

8.1 Formal vs. informal evaluations

Although ADT is grounded in mathematical statistics, the tools for the lay decider are in the form of simple numerical models, using no more than elementary arithmetic. Moreover, they will not replace your regular decision processes; only complement them — and then not always. However, although the models themselves are simple, the thinking that goes into using them can be complex. Moreover, learning to work with these models should educate you to apply qualitative procedures that are based on them.

ADT is becoming widely used in business, government and medicine, where the stakes are high enough to justify the trouble of doing it right. Its practical impact in private decision-making is quite limited, but practice doing quantified ADT, in a course based on this text, can enhance your informal rationality and intuition, for example in the form of verbal guidelines.

Most of the contribution ADT can make is usually achieved with half a dozen very spare and structurally simple modeling tools. This is not to say that effective ADT is easy; far from it. But most of the art comes from knowing how to use tools cost-effectively (that is, producing maximum benefit with minimum effort) in a way that brings an improvement over existing practice.

8.2 Caution

There is a risk of over-relying on ADT and other quantitative evaluations. An *ideal* analysis of everything you know on an issue, not necessarily explicit, will very likely outperform your unaided judgment. However, no real analysis ever manages to be ideal, and any particular analysis may fall so far short that its results are actually less sound than your unaided judgment.

♦ A short introductory text on ADT for managers that I coauthored[1] was used in executive training programs. A frustrating but common aftermath of this course was that students who mastered the basics felt that they knew how to do decision analysis effectively. Then they would come back a few months later to say, "I tried it, but it doesn't work." I hadn't made sufficiently clear to them that in a two-week course they could only expect to appreciate principles, not to apply them successfully to live decisions.

♦ A colleague found the same thing by assigning students to do a decision analysis in a course on judgments and decisions, after a couple of textbook chapters and a few lectures. The students made obvious mistakes, such as forgetting to include obvious outcomes, or including, as outcomes, their own initial preferences for options, or counting tiny differences between some outcomes of options as if they were much more important than they were. As a result, the students often found the results either uninformative (agreeing with what they already thought, because they included that as an outcome) or completely nonsensical (as they were).

Quantifying a decision evaluation may inspire more confidence than it deserves. There are potential pitfalls in the use of ADT (or other quantitative modeling approaches). It may divert attention from the substance of an argument, to the technique for handling it. The risk of mis-modeling is great and can lead to errors much worse than your unaided intuition might.

It is dangerous to accept a plausible quantitative argument, such as ADT, without robust safeguards (such as a common-sense reality check).

♦ *Judicial case.* Throughout 1995, TV viewers were regaled by "the trial of the century." Despite damaging evidence, sports star O.J. Simpson was acquitted of murdering his wife and a friend. Imagine that Simpson had in fact confessed and defended his murder decision with the following argument: "When I decided on the murders, I scientifically weighed the pros and cons. I

---

[1] Brown et al., 1974.

took into account the value of a human life, which environmental regulations have put at $20 million, so the social cost of my murders was $40 million. I accurately predicted that I would go on trial, and generate public entertainment worth $10 each to 100 million TV viewers. Thus, I made a decision that promised a social return of $1 billion on an investment of a mere $40 million. I deserve commendation for my spirited public choice, not punishment."

This example — admittedly twisted to make a point — illustrates a common situation, where judgments concerning the same choice conflict. To reconcile that incoherence, you must revise at least one of the judgments and reconsider the judgments they are based on — *somehow*.

The unrealism in this highly hypothetical case is easily caught, because it is so powerfully counter-intuitive. In less obvious cases, an unsuspecting decider may accept and act upon a flawed model with harmful results.

8.3 Combining an ADT analysis with alternative evaluations

The major flaw in much ADT practice, responsible for many results that fail to outperform your unaided judgment, is that the decision models do not take into account much valuable knowledge that you have access to and already use implicitly in your regular thinking.

The often simple, but rarely used, device of *hybrid reasoning* is usually sufficient to avoid the most harmful failures of flawed ADT. Hybrid reasoning is addressing a judgment several different ways, and then merging them — *somehow*.

♦ Joe's decision tree model in Figure 3 implies that he should choose Lulu over Jane (71 > 65.5). However, say, his initial holistic judgment had been to evaluate Lulu as having a utility of 55 (on the same utility scale), so that Jane would now be preferred to Lulu. How do you resolve this inconsistency? Joe, on reflection, feels that his holistic judgment had been clouded by Lulu's appearance, and that he drastically over-estimated his utility in the situation where Lulu would not get a good job. So he revises that utility downward, from 55 to 15. This reduces Lulu's utility to 63, which is now a bit less than Jane's utility. The two approaches now cohere, although the choice is still close. So Joe switches his preference, from Lulu to Jane.

Chapter 9

A civic case: Voting for president

9.1 Leah's vote

Leah is undecided about whether to vote for Romney or Obama in the 2008 presidential elections. She has been impressed by Romney's performance in a TV debate and is leaning in his favor, although both candidates' arguments have left her confused on substance. She wants to use what she has learned in a short course on applied decision theory to make the wisest possible choice, based on everything she knows and feels.

She evaluates the candidates on various criteria, using a simple pro-con tallying approach (Table 4). She marks a perceived advantage to Obama as one or more pluses and an advantage to Romney as one or more minuses.

Leah took into account both the importance of each criterion and the candidate's impact on it. (She decided against quantifying the two components, due to commonly occurring elicitation difficulties.)

Other presidential attributes were important to Leah, but they did not differentiate the candidates and so did not require an entry to the tally. In particular, a president's contribution to the national economy was important to her, but did not favor one candidate over the other. Although she expected Romney would make a somewhat greater contribution to the economy, it was also more uncertain than Obama's, which canceled Romney's perceived advantage in that respect.

Table 4: Leah's Tally: Obama − Romney

CHARACTER	
Experience	+
Political effectiveness	0
Judgment	+ +
Values	0
Ethnicity	+
Personability	+
	+5
POLICY	
Long-term Deficit	+
Short-term Economy	+
Support for Israel	− − − − − −
Health care	+
Social equity	+
Other factors	0
	− 2
NET:	+3
PREFERENCE	OBAMA

Tallying these pluses and minuses has her preferring Obama, which contradicts her initial inclination to vote for Romney. Leah seeks to reconcile these two conflicting indications so that they both reflect all she knows and feels.

She is led to conclude that:

1. Her tallying method does not allow for effects smaller than a single plus or minus. Several of Obama's positive evaluations merited less than a full plus.

2. Her initial strong preference for Romney was biased by pressure from her pro-Israel friends, who favor Romney as more supportive of Israeli government actions.

9.1. LEAH'S VOTE

3. Her tally double-counts some of Obama's positive evaluations. For example, improvements in health care should also benefit the deficit and the economy.

Accordingly, she redoes her tally with numbers, treating one plus as ten points, etc.

Table 5: Leah's adjusted numerical tally Comparison Obama – Romney

CHARACTER	
Experience	+8
Political effectiveness	−1
Judgment	+17
Values	+9
Ethnicity	+1
Personability	+4
	+38
POLICY	
Long-term Deficit	+12
Short-term Economy	+7
Support for Israel	−60
Health care	+3
Social equity	+10
Other factors	0
	−28
Double-counting adjustment	−15
NET PREFERENCE ROMNEY	−5

Now Romney comes out slightly ahead, which confirms her initial preference for him. However, she also recognizes that her preference was exaggerated by pressure from her strongly pro-Israel friends. Thus, adjusting both tallying and intuitive approaches in opposite directions reconciles the results.

Leah's criterion evaluations combine both factual and value judgments. (She could have elaborated her analysis further by decomposing evaluations into factual and value components — e.g., as the product of a criterion score and an importance weight. However, she understands that it is cognitively difficult to elicit weights that correspond to the scoring scales used, without more skill and effort than she has available.) A high evaluation could be because either she judges the difference between candidates to be great or the criterion is very important. For example, although she estimates that Obama's experience is only slightly greater than Romney's, she attaches great importance to this criterion, so its contribution to Obama's net evaluation is appreciable.

Leah acknowledges that her judgment could be ill-informed. For example, the current Israeli policy on West Bank settlements could actually be *detrimental* to Israel's interests (which accounts for Abe's contrary evaluation on that issue, in our next example). Such a possibility could lead her to regret having voted for Romney. She could do research to resolve such uncertainties. She feels it her civic duty to make the best voting decision, but since her single vote will have negligible effect on election results, she does not bother with the extra reading needed.

9.2 Abe's vote

She does, however, confer with her classmate Abe, who strongly supports Obama. She has him make his own tally in the same format as hers.

Thus, Abe's exercise strongly favors Obama, based on both his initial inclination and tally. He is confident that further deliberation would not change his mind and so it would not be worth him spending more time on the matter. He will certainly vote for Obama.

Exploring the rationales behind their conflicting evaluations, Leah notes that they do not differ significantly on *factual* judgments. In particular, they agree that Obama is likely to stand up to the Israeli lobby (since he does now have to worry about re-election) and will try to influence Israel's West Bank settlement actions. So their discussion only serves to reinforce Leah's factual judgments.

9.2. ABE'S VOTE

Table 6: Abe's numerical tally comparison.

CHARACTER	
Experience	8
Political effectiveness	−1
Intellect	17
Integrity	9
Personal appeal	4
Ethnicity	1
	38
POLICY	
Long-term Deficit	8
Short-term Economy	7
Support for Israel	2
Health care	3
Social equity	5
Other factors	20
	45
Double-counting adjustment	−12
NET	71
PREFERENCE	**OBAMA**

Where they differ is on how they *value* the criteria. For example, they agree that Romney would support the Israeli government more, but differ on whether they approve or disapprove of that.

Such value judgments are somewhat fixed. (*De gustibus non est disputandum* — tastes cannot be disputed.) Leah's values are not much influenced by knowing Abe's values. Note that values are only "somewhat" fixed because criterion importance has a factual component. This is vulnerable to new evidence, which a value judgment is not. Importance weights *predict* ultimate utility. Leah weighs candidates' support for Israel heavily because she believes *factually* that the support will enhance Israel's security which is what she really cares about.

Although her considered (but, she admits, fallible) judgment favors Romney, at the last minute, a scandal erupts that casts Romney's private behavior in an unsavory light. Although Leah does not believe that bears in any way on his fitness as president, she develops a sudden distaste for him personally and on impulse votes for Obama (which validates the saying "Thinking smart is not the same as acting smart"). Leah may have come to be glad she voted for Obama. Nevertheless, voting for Romney may still have been her wisest choice given what was in her head at the time. Her thinking was quite rational.

9.3 Exercise

Pick any two-person election (preferably current) that you are interested in. Evaluate how you would vote, expressing your reasoning in the same form as Leah's. If possible, check it against the judgment of someone you know who is disposed to vote the other way. How much of your disagreement is due to differences in your values as opposed to differences in your factual judgments?

Chapter 10

Information value case: Life-saving diagnosis

I credit ADT (applied decision theory), and Bayesian updating in particular, for guiding my doctor to diagnose me with pancreatic cancer in time for successful surgery, in 2011. Without it, I would almost certainly not have lived to tell this tale.

10.1 Decision strategy

In mid-2011, I reported to my physician — we'll call him Dr. D — that I had been losing a great deal of weight during at least the past year. I was concerned, since I understood that weight loss is common with some types of cancer. D assured me that a weight drop was not too surprising, given the strict diet he had put me on for my diabetes. He advised "watchful waiting".

I was not entirely reassured. It was clear that it was in *my* best interest to get tested for cancer. It was essentially free to me, since insurance would bear the cost, but D would responsibly want to take cost into account. I had no wish to use more than my fair share of scarce testing resources. The question was, should D judge my chance of having cancer high enough above the general population rate to warrant testing?

There were two issues to address: what is the likelihood that I have cancer? Is it high enough to warrant testing? I felt, as is commonly the case, that diagnostic effort analyzing the former promised to be more productive than the latter. Accordingly I devoted most analytic effort to diagnosis, using *Bayesian updating*, which has an excellent "track record". My analysis of D's reflections (see the following) enabled me to persuade D that I should be tested. I was. Pancreatic cancer stage two was diagnosed. This diagnosis was early enough to permit successful surgery, which put the cancer in remission. Without the surgery, I would almost certainly have died within the year.

10.2 Diagnosis

I was conscious of having more symptoms than the simple weight loss that I had told D of. In particular, I had recently been losing weight *at a regular pace* of some three pounds a month, in fact down some 55 lb. over the past couple of years, from 185 lb to 135 lb, which my eating habits seemed unlikely to account for. Moreover, I had noticed other surprising clues that that I could not adequately communicate to D. They included a mild but consistent discomfort around my midriff — and an unfamiliar general and also persistent and distinctive unwell feeling.

As a layman, I hesitated to take issue with D, a well-regarded physician. However, I thought perhaps I could help him make a responsible testing choice, drawing on his own medical expertise and my logical (ADT) perspective.

I asked D to focus on three questions, whose answers were needed for a "Bayesian diagnosis" under two information conditions: the original minimal weight-loss symptom alone (a); and the richer maximal symptoms I was conscious of (b). The following gives the gist of our exchange.

1. Prior probability of cancer (C). R: "Before learning of any symptoms, what chances do you give that I have cancer?" D: "Somewhat higher than in the population at large, whatever that may be, taking into account your advanced age (76)."

2. Probability of symptoms (S), conditional on C. R: "If I do have cancer, how surprised would you be at learning of these symptoms?" a) Absolutely unsurprised; b) slightly surprised.

3. Ditto, conditional on *no* cancer ($\sim$C). R: "And if I *don't* have cancer?" D: a) "Somewhat surprised"; b) "most surprised".

10.3. ASSESSMENTS IMPUTED TO D'S DIAGNOSIS

4. R: "On the basis of your intuition, what would you prescribe?" D: "a) Watchful waiting, just short of justifying testing; b) order testing".

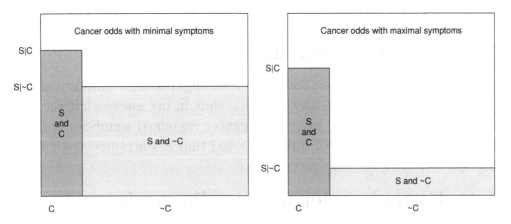

Figure 4: Analysis of cancer risk given minimal (a) and maximal (b) symptoms.

10.3 Assessments imputed to D's diagnosis

The reasoning which enabled me to change D's prescription from watchful waiting to life-saving surgery is captured in Figure 4. From the previous discussion, I cautiously imputed to D quantitative (but not numerical), inputs called (C) for in Bayesian diagnosis where a) evidence consisted of the initial weight-loss only symptoms and b) the more detailed symptoms that I reported later.

The horizontal axis shows, by eye, a prior probability of cancer (C) of about 20%, which is the same for both symptom cases. The vertical axis shows the conditional probability of the symptoms in each case, given cancer (S|C) and no cancer (S|~C), for both cases. The areas of the boxes thus formed give "joint" probabilities of symptoms and cancer (C and S), and of symptoms and no cancer (~C and S).

Comparison of those two areas gives the relative probabilities, or odds, of cancer given symptoms (C|S) — which is our target inference. With the assessments shown in Figure 4, comparison of areas shows probability of cancer increasing from about 25% to 54%, with the more detailed symptoms.

I inferred from this that D's initial advice of "watchful waiting" that my risk of cancer appeared then to be just short of warranting testing. Consideration of the more detailed symptoms more than doubled that threshold risk and clearly warranted testing.

10.3.1 Sensitivity analysis

Recall that the quantities embodied in these figures were only my *guesses* at D's judgments. However, any plausible shift in my guesses left the main message intact — i.e., that my risk of cancer remained significantly higher than what indicated watchful waiting — and that it warranted testing.

10.4 Formal underpinning diagnosis

The analysis underpinning my own thinking was informal and number-free, though implicitly consistent with quantitative ADT, in particular "Bayesian" diagnostic updating. (In my opinion, to express the analysis in numbers would have been unnecessarily time burdensome and difficult to communicate.)

Figure 4 shows graphically a quantitative (but still number-free) ADT representation of the earlier reasoning. It is more precise, but in this case more precision was not needed. The earlier informal analysis was sufficient to satisfy D that testing was warranted. The rectangles (which I call "Bayes boxes") correspond to assessed cancer probabilities, corresponding to D's assessments with only the weight-loss symptoms (a) and with all available symptoms (b).

The bases (widths) of the boxes correspond to the "prior" probabilities of cancer (i.e., without taking *any* symptoms into account). The height of the boxes corresponds to the diagnosticity of the symptoms, i.e., the probability of the symptoms conditional on whether I have (S|C) or do not have (S|∼C) cancer respectively. The relative box areas give cancer odds (according to ADT Bayesian logic) and therefore probabilities. The figures have the same box bases (since prior probability is not affected by symptoms).

It can be judged by eye that the cancer rectangle (C) on the left of Figure 4 has about a third of the area as the no-cancer (∼C) rectangle, indicating that the odds of cancer are about 1:3, or 25% probability. Those minimal symptoms led to D prescribing "watchful waiting", which I took to mean that the threshold probability, for testing was just above 25%. The right figure shows cancer odds (C vs. ~C) with maximal symptoms to be about equal, or about a 50% probability of cancer (actually 54% in the figure). The rectangles have roughly equal area.

10.5 Does diagnosis warrant testing?

Setting a hurdle for cancer probability — for testing more analytically — would implicitly take into account a complex set of factual and value judgments. Theoretically, this is a complex ADT "value of information" exercise. In these circumstances, that would be practically and cognitively too burdensome and error-prone to do explicitly. Here, D only had to understand a hurdle probability precisely enough to say, with some confidence, whether or not his actual cancer probability exceeded it, and so warranted testing. An experienced and all-wise doctor could have taken these considerations properly into account, perhaps subconsciously[1], and act according to an appropriate hurdle cancer probability. D appears to have cleared this hurdle, in this case.

Defining a hurdle probability more precisely would depend on *whose* interests are to be served. For me, the hurdle was trivially low, since my insurance covered the cost of testing. But the testing decision was to be made by D, who, as a responsible professional, would respond to broader social concerns. Testing resources are limited, so he would need to make sure that testing me does not come at the expense of someone more deserving. Saving the life of a 76-year-old gains fewer years of life than someone younger. So D's hurdle for cancer probability would be higher than mine.

As argued earlier, cancer risk with the maximal symptoms appeared to be at least twice as high than with minimal symptoms (which indicated "watchful waiting") and would intuitively clearly warrant testing.

---

[1]I am not sure that D fully followed my ADT reasoning, but was swayed by academic credentials. If so, I suppose a Harvard doctorate in necromancy would have served just as well as in decision theory!

10.6 Credit to ADT?

I attribute my initial lucky escape to ADT, in the sense that it would not have happened without my drawing on ADT logic, however loosely. I would not have been tested and the cancer would have advanced to an inoperable stage. I did not need ADT to decide that testing was in *my own* best interest, since that was easy to establish. The cost of surgery would be largely born by my insurance, which clinched the case for me. Thus, the cost to me was insignificant and the potential benefit immense. D's choice was a less clear case.

However, if my intervention in D's decision process resulted in enhancing his choice — to test — it may have less to do with ADT methods, and more to the fact that I had more diagnostic evidence I used (albeit in an ADT framework). D's initial no-test advice was based on limited weight-loss evidence, which was not enough to warrant testing. I had a much richer, more diagnostic knowledge of my symptoms. If D had had such information (which he would have if he were diagnosing his *own* condition), perhaps, he would have made at least as sound an inference about my cancer, *without ADT*. He may well be a competent "intuitive statistician", but ADT enabled me to focus attention on critical judgments, notably on the diagnosticity of my symptoms. I would like to think my superior ADT inference logic gave me an edge, but I cannot be sure.

Be that as it may, I submit that I did need ADT to persuade my doctor that testing was warranted and to act on it. Without ADT, I would not have had the *confidence* to argue for testing forcefully and persistently. Moreover, without ADT, I doubt that D would have been *persuaded* to change his mind on testing. ADT may have provided a persuasive *rationale* for testing. (As for the biopsy choice, I doubt that my ADT training benefited me much, if at all. It didn't appear to give me an edge over the soundness of the deliberations of family members.)

10.7 Assignments

1. Suppose Dr. D accepts as reasonable a hospital guideline that doctors generally test for pancreatic cancer only if they assess more than 50% chance that the patient has it.

    a. If D adopts the assessments in Figure 4, would he test me?

    b. If a guideline is changed to testing only if cancer assessment is 25% or more should I now be tested?

2. Express the biopsy choice as a plus-minus tally exercise, making plausible guesses at any needed judgments.

10.8 Postscript

Although I was given several more years of life, in October 2016 a CT scan showed that "recurrence of cancer is very likely." I had to decide whether or not to have a biopsy which would indicate, with certainty, whether the cancer had indeed returned. If it had, no further medical treatment such as chemotherapy was feasible. My oncologist, Physician P, advised me that the only benefit from a biopsy would be to satisfy my curiosity and it would use up some hospital resources. I thought that knowing I had cancer would very likely change my behavior, but I could not think exactly how. P's attitude weakly suggested mild discouragement of biopsy.

I judged that knowing my condition for sure would be worth a little, but not the burden of a biopsy, which included my personal inconvenience and other unidentified health risks. I took some account of social cost — the diversion of scarce testing resources, which could be used for the greater benefit of someone younger, with many more years to enjoy. I decided against the biopsy. Family members agreed with me. I attended a pre-surgery consultation which led to nothing to change my mind. I canceled surgery and decided to let nature run its course.

Chapter 11

Expanded view: Should we teach decision making in school?

11.1 Reading, writing, arithmetic... and decision making?

For years, educators have been calling for more attention in our public schools on "critical thinking skills". And yet the one skill which could most improve our lives, decision-making, has made little headway.

That doesn't have to be. Wise decision-making is a teachable skill that is now developed to the point where it could be a central part of everyone's school experience. A revolution in educational practice is needed to make that happen, but that will not easily come about.

Few would dispute the need to improve decisions throughout society. Nearly all of us, young or old, make unwise decisions all the time — and pay for them dearly. Through habit, laziness or confusion, we take action that we could greatly improve on. That is, we could reasonably expect that some other action would increase our satisfaction.

We stay in a dead-end career, marry someone who is patently unsuitable, squander our retirement savings or fail to have life-saving surgery. We vote for politicians who advance special interests and neglect ours or the world's. We make poor management and professional decisions in the work-place, which hurt ourselves and the economy.

It is certainly no small challenge to improve on decision skills shaped by natural selection. Cave-men who made wise choices between fight and flight tended to survive. However, industrial and other revolutions since then have diversified the dilemmas we face, and we need new decision skills to survive.

A revolution in school education is no longer held up by lack of appropriate teaching materials. (The obstacles are institutional and political, but I will come to that.) A richly-stocked and well-established kit of decision-aiding tools is now available. They are currently widely taught in universities and used in business, government and medicine, and variants are available that are simple enough to fit school-children's needs and capabilities

At its most basic, decision skill is just sharpened common sense. For example, the decider asks himself three universal questions. "What can I do? What will happen? How will I like it?" An imaginary voter might reason: "I could support tougher gun controls. They would save a few lives, but cost us constitutional rights and set a dangerous precedent. The last two issues outweigh the first for me." Based on these very personal answers, the voter concludes that he should oppose gun control.

Another readily accessible decision tool is alerting deciders to common psychological biases and providing guidance on how to avoid them. For example, many people have the unfounded conviction that past trends will persist indefinitely. A few years ago, countless homeowners were certain that soaring house prices would continue to rise. They bought property they could not afford and lost their shirts when the housing bubble burst. Had they been mindful of their bias, they could have corrected for it and avoided disaster.

A variety of powerful tools are available to more advanced students. Decision reasoning can even be turned into numbers (for example, with probabilities and "decision trees"), but in real life people rarely have the time or training to benefit. Nevertheless, training in quantitative decision analysis life has a significant role to play in educating intuition and informal reasoning. Former MBA students and faculty colleagues who have become powerful executives tell me that much of their success is due to their decision theory training, though they almost never quantify their decisions. There is something to the saying that education is what you have left when you have forgotten everything you learned.

I claim that I owe some of my own life to informal use of my decision training. As described in Chapter 10, my own analysis clearly indicated I should get tested for cancer when the doctor did not think that was necessary. My argument persuaded the doctor and he authorized testing. The result was early treatment, without which I would not have been able to write this book.

Curricula are now available to teach decision tools at several levels of middle and high school and have been tested in pilot courses. They have got students to practice the skills on real dilemmas that face them individually, such as whether to drop out of school or go to college.

So why aren't decision skills more widely taught in school yet?

In the first place, promising though teaching experiments have been, there is not yet convincing evidence that an extensive school program can improve students' life decisions enough to justify a major national commitment. It is not easy to demonstrate that students who are trained in decision-making lead more satisfying lives than those who aren't.

The tepid response to past calls for decision and other critical thinking skills has so far consisted mainly of regular teachers inserting material into their traditional school courses, such as English or History. The focus has been on improving students' course performance, rather than their thinking outside the class and making wiser decisions. That could require offering entirely separate courses, with specialized instructors not distracted by teaching another subject.

Such instructors will require extensive and distinctive training and likely be chronically in short supply. A teacher qualified to improve on students' natural decision-making needs a rare blend of quantitative logic, human performance, practical realism and the ability to adapt teaching to a student's individual experience.

The material is intellectually challenging and teachers, let alone students, often find the technical content of decision tools daunting. One student commented "My other courses are 80% absorption and 20% thinking; yours is the other way round." I know of no institution of higher education where a would-be decision teacher can get the balanced, interdisciplinary training needed, which can assure him a secure career, where he can advance without having to satisfy colleagues in another discipline.

The main impediment to teaching both personal and civic decision-making in school is social and political resistance. In particular, teachers' unions are protective of members whose own subjects may be displaced. Furthermore, people of many kinds are concerned lest youths who learn to make their minds up independently will come to the "wrong", i.e. inconvenient, conclusions. These people include parents, politicians, activists, religionists and even some doctors and educators.

On the other hand, as voters, few citizens are motivated to support wise legislation. A colleague, with whom I had taught decision skills to MBAs, had become a highly successful businessman and was in the process of running for the US Senate. We discussed possible platforms for his campaign. I suggested: "Why not put yourself forward as the champion of logical decision making in government? You certainly have the credentials ". He replied "You don't understand. Voters don't care if you make logical decisions for them. People would sooner watch a fight than listen to a lecture. And the faint prospect that their single vote will make a difference is swamped by other considerations, like feeling good. Do you have any ideas I can get into 30 seconds of TV?"

On the other hand, I have found that the general public is unreservedly enthusiastic about decision training for personal (rather than civic) decisions. So are many students, in particular, "disadvantaged" youths. They face dilemmas, where bad choices may ruin their lives, involving truancy, criminal behavior, birth control and drugs.

One of our students told us "At last the establishment is teaching us something we can use."

Proceeding initially in small steps may forestall the most discouraging of these obstacles. Get the camel's nose under the tent first. A promising first step could be to open an office of decision making at the Department of Education, which could focus initially on supporting underfunded private initiatives, such as the Decision Education Foundation. If early progress were encouraging, this office could develop into a dedicated institute, perhaps comparable to the National Institute of Child Health and Human Development. Analogously, acupuncture was initially promoted inconspicuously by an "Office of Alternative Medicines" at the Department of Health, and is now well on the way to becoming maim-stream medicine.

The task of a decision institute would extend to higher education, where it already has a firm, if haphazard, foothold. Many business schools, for example, teach applied decision theory and the psychology of decision bias. Adaptation of such practice for grade school may be the best entry point.

Attempts at universal decision education are not without precedent. In the 1980s, a newly created Venezuelan Ministry of Intelligence mandated that decision skills be taught throughout the country's public school system. However, the experiment was discontinued after six years, when the "Minister of Intelligence" driving the project left office. Its short life may be largely due to trying to do too much, too soon and to the immaturity of the decision tools then available.

Anthony Bryk, President of the Carnegie Foundation for the Advancement of Teaching, has observed that, as a fraction of total budget, education spends less than 5% of what other fields do on research and development. So adequate funding for a decision training program ought not to be an insuperable problem — if the support of education authorities can be enlisted.

Make no mistake. It will take years and generous financing to do the course development, teacher training, pilot testing and performance evaluation needed before anything like fully functioning decision training is in place throughout the public school system. The focus of educational reform has so far been almost exclusively on *how to teach* existing curricula of math, science and literacy. The national interest would be better served if they also pay serious attention to *what is taught*.

11.2 Exercises

11.2.1 Initial assignment

Whatever your actual opinion, compose a letter, of ten lines or less that a reasonable reader who *disagrees* with this proposal might write in response.

11.2.2 Exercises for end of course

Summarize and recast this article in the form of a term paper, without substituting judgments of your own, as far as possible.

Chapter 12

Epilogue — What next?

12.1 Making use of what you have learned

You may rarely, if ever, find a numerical analysis of a real dilemma cost-effective.

However, if the choice is one-step and the stakes are high, and you would do well to use informal ADT, in particular, "go through the HOOPS", or sketch out a decision tree to structure your unfolding thinking.

Probably most of your consequential decisions will be (and should be) a complex progression of "muddling through". Your ideal sequence of thoughts and actions through this process would conform to ADT logic, but implicitly, and what you have learned here should, consciously or subconsciously, steer you usefully toward it.

12.1.1 Basic ADT logic

This text has presented only a brief introduction to the principles and tools of applied decision theory and its role within more general prescriptive decision analysis. However, if you have occasion to put these tools into practice in your day-to-day decisions, you may have absorbed reasoning habits that will immediately nudge you toward greater wisdom and satisfaction, especially if you can get feedback from someone more experienced in applying these tools. It has been said that education is what you have left when you have forgotten everything you learned. To some extent, that is true of the decision training you have been exposed to. In particular, you will be more skilled at knowing what questions to ask and what to do with the answers, informally or intuitively.

If you have worked your way diligently through the text and assignments, I hope you'll be in position to make wiser choices and other judgments — and be poised to get progressively better, as you practice from day to day what you have learned.

In general, I hope you will benefit from having your intuition and informal reasoning educated by the ADT perspective, even if you do not use any tool explicitly. When you face the straightforward dilemma of choosing among a few options, the ADT structuring concepts should be helpful.

I do not want to oversell the usefulness of what you have learned here. Your regular decision-making reasoning, often interleaving thought, inquiry and primary action, may not change much. However, in principle, every step of the way, your thinking should *conform* to a theoretical ideal hidden inside your head. In this text, we have explored just one avenue to approaching that ideal. That is to compare a few options, as if you were committing to one of them once and for all. Although this not usually a realistic assumption, I believe that learning how to do it effectively helps you proceed along an appropriate, more complex path, which comes as close as you can get to your ideal decision strategy. A significant exception is the use of Bayesian diagnosis (Chapter 9) to update your uncertain assessments in the light of new information. In particular, when you are faced with a major personal dilemma, you will do well to follow the normal complex progression thought process, which does not lend itself well to explicit ADT structuring. However, when you do it well you will (perhaps unknowingly) be close to your ideal judgment according to decision theory norms.

12.2 Developing your reasoning further

This text is designed to be self-contained, but there are several productive further ADT directions for the eager student to go in. There are various ways to build on this text:

- A highly regarded introduction to ADT by leading authorities in the field (Hammond et al., 1999) may usefully complement this text, with

12.2. DEVELOPING YOUR REASONING FURTHER

little unnecessary overlap. In particular, it presents "even swap", an innovative way of dealing with conflicting objectives.

- Keeney (1992) provides a similar approach, with emphasis on values, including many examples of the sort discussed in this text. The middle section of the book is mathematical and somewhat independent of the rest of it.

- Several books cover decision analysis in greater formal and theoretical depth (e.g., Watson & Buede, 1987) and introduce advanced analytical techniques (e.g., Clemen, 1996).

- Other books review the role of decision analysis in specific fields such as medicine. Chapman and Sonnenberg's (2000) edited book provides a broad perspective on that field.

- Various scholarly societies concern themselves with decision analysis, primarily the Decision Analysis Society and the Society for Medical Decision Making, but also (less so) the Society for Judgment and Decision Making and the European Association for Decision Making.

- Several academic journals (some associated with these societies) publish papers about decision analysis, especially Decision Analysis, and Medical Decision Making.

- The psychology of decision making is covered in many other books. Baron (2008) covers this and also includes a couple of chapters on decision analysis itself, very much in the spirit of this text. Kahneman (2011) provides a popular introduction to his own important work. Kahneman and Tversky (2000) collected some classic articles.

I maintain that developing this elementary introduction to ADT into more ambitious expertise can confer major benefits to your life, and possibly to the lives of others.

Bibliography

Baron, J., & Brown, R. V. (Eds.). (1991). *Teaching decision making to adolescents.* Hillsdale, NJ: Erlbaum.

Baron, J. (2008). *Thinking and deciding* (4th ed.). New York: Cambridge University Press.

Brown, R. V. (2012). Decision theory as an aid to private choice. *Judgment and Decision Making, 7*, 207–223.

Brown, R. V., Kahr, A. S., & Peterson, C. R. (1974). *Decision analysis for the manager.* New York: Holt, Rinehart, & Winston.

Brown, R. V., & Ulvila, J. W. (1982). Decision analysis comes of age. *Harvard Business Review, 60*, 130–141.

Chapman, G. B., & Sonnenberg, F. A. (2000). *Decision making in health care: Theory, psychology, and applications.* New York: Cambridge University Press.

Clemen, R. (1996). *Making hard decisions: An introduction to decision analysis.* (2nd ed.) Belmont, CA: Duxbury.

Hammond, J. S., Keeney, R. L. & Raiffa, H. (1999). *Smart choices: Making better decisions.* Boston, MA: Harvard Business School Press.

Kahneman, D. (2011). *Thinking, fast and slow.* New York: Farrar, Strauss and Giroux.

Kahneman, D., & Tversky, A. (2000). *Choices, values, and frames.* New York: Cambridge University Press (and Russell Sage Foundation).

Keeney, R. L. (1992). *Value-focused thinking: A path to creative decision-making.* Cambridge, MA: Harvard University Press.

Raiffa, H., & Schlaifer, R. (1961). *Applied statistical decision theory.* Harvard University Press.

Watson, S., & Buede, D. (1987). *Decision synthesis.* New York: Cambridge University Press.

Index

ADT, 1, 4, 31, 68, 95
    alternatives, 73
    benefit, 86
    over-reliance on, 72
    quantitative, 53, 71
anchoring judgments, 33
applied decision theory, *see* ADT

caesarean section, 61
cancer, 16, 81
child-birth, 61

decision rules, 34
decision trees, 56
decision-analysis, 14
decisions, 9
    civic, 19
    personal, 19
    professional, 20
    vs. action, 9, 17, 50
    vs. outcomes, 22

expected utility, 2
expert opinion, 34

hip surgery, 43
holistic judgment, 35
HOOPS, 35, 36, 39
    quantifying, 56

ideal judgment, 15

information, 49
intuition, 44

Jane, *see* Joe
Joe, 13, 30, 32–34, 55, 57

knowledge, 12

Lulu, *see* Joe

mind-contents, 17
models, 53, 71

options
    simplifying, 46
outcomes, 35

precepts, 18
preferences, 35
prescriptive decision analysis, 29
process, 18
project, 26, 38, 45
projection, 31

qualitative decision aids, 29

rationality, 16
reference comparison, 31

sensitivity analysis, 84
statistical decision theory, 5
synthesis, 36, 45, 48

teaching decision making, 89

utility, 2, 55

values, 35
    vs. facts, 36, 41
visualizing outcomes, 31, 45
voting, 16, 75

wisdom, 10, 11, 17